CHASING NORTH AMERICAN MONSTERS

About the Author

© Photo by Jason Offutt

My most impressionable years were during the Sasquatch sweet spot, the 1970s. This was well before Hollywood gave us the family-friendly *Harry and the Hendersons* (1987), starring John Lithgow and Don Ameche, but well after the surprisingly respectful *The Abominable Snowman* (1957), starring Peter Cushing and Forrest Tucker (yes, *that* Forrest Tucker. The guy from *F Troop*. The one with the funny hat).

But the 1970s was *the decade* for Bigfoot. Check out some of the films:

- *Bigfoot* (1970)
- *The Legend of Boggy Creek* (1972)
- *The Legend of Bigfoot* (1975)
- *Curse of Bigfoot* (1975)
- *Creature from Black Lake* (1976)
- *Sasquatch: The Legend of Bigfoot* (1976)
- *Snowbeast* (1977)
- *Return to Boggy Creek* (1977)
- *The Capture of Bigfoot* (1979)

Are these movies complete shlock? Of course, but they're awesome shlock. Imagine pulling into the drive-in theater in your Gran Torino just to watch *Return to Boggy Creek*. Of course, you're dressed in bell-bottoms and a Keep On Truckin' T-shirt, but at that time, who wasn't?

I was hooked on monsters then, and I still am. I've written one book on cryptozoology, *Chasing American Monsters* (2019), four books on ghosty things, and a number of novels about the spooky and paranormal, like the award-winning horror novel *The Girl in the Corn* (2022) and its sequel, *The Boy from Two Worlds* (2024).

I teach university journalism, cook for my family, drink lots and lots of beer, and spend too much time wondering how I could use our cat to ward off a werewolf attack. Sorry if that ever happens, Gary. We'll all remember you as a good boy.

You can find more about me at www.jasonoffutt.com. And if you see me on social media, please tell me to get back to work.

CHASING NORTH AMERICAN MONSTERS

A Guide to Over 250 Creatures from Greenland to Guatemala

JASON OFFUTT

WOODBURY, MINNESOTA

FIRST EDITION
First Printing, 2025

Book design by Samantha Peterson
Cover art by Michael Koelsch
Cover design by Kevin R. Brown
Interior illustrations by Llewellyn Art Department

Library of Congress Cataloging-in-Publication Data (Pending)
ISBN: 978-0-7387-7893-8

Llewellyn Publications
A Division of Llewellyn Worldwide Ltd.
2143 Wooddale Drive
Woodbury, MN 55125-2989
www.llewellyn.com

Printed in the United States of America

GPSR Representation:
UPI-2M PLUS d.o.o., Medulićeva 20, 10000 Zagreb, Croatia,
matt.parsons@upi2mbooks.hr

Other Books by Jason Offutt

Paranormal Nonfiction

Chasing American Monsters: 251 Creatures, Cryptids, and Hairy Beasts (2019)

What Lurks Beyond: The Paranormal in Your Backyard (2010)

Paranormal Missouri: Show Me Your Monsters (2010)

Darkness Walks: The Shadow People Among Us (2009)

Haunted Missouri: A Ghostly Guide to the Show-Me State's Most Spirited Spots (2007)

Fiction

The Boy from Two Worlds (2024)

The Girl in the Corn (2022)

So You Had to Build a Time Machine (2020)

Bad Day for a Road Trip (2018)

Bad Day for the Apocalypse (2017)

Road Closed: Twelve Bloody Stories to Brighten Your Day (2017)

Matriarchal Nazi Cannibals (2016)

A Funeral Story (2014)

Humor

How to Kill Monsters Using Common Household Items (2015)

Across A Corn-Swept Land: An Epic Beer Run Through the Upper Midwest (2013)

On Being Dad (2005)

Biography for Elementary Readers

J. C. Penney: The Man with a Thousand Partners (2017)

Ella Ewing: The Missouri Giantess (2016)

Helen Stephens: The Fulton Flash (2014)

Forthcoming Books by Jason Offutt

Chasing the World's Monsters (2026)

For my children, who are the only
monsters I've ever needed.

Contents

Foreword
Chasing Monsters Through Romantic Zoology and Cryptozoology

Five years ago, I highly recommended Jason Offutt's first overview of American cryptids by writing he did "a special service to the field of cryptozoology...by keeping all of us up to date and incredibly informed—beyond the scope of lesser guidebooks."

Now we are revisited by Jason's insights in a new and varied menagerie of critters. But how did we find ourselves here? How did cryptozoology evolve into our lives?

When I first began reading and writing on the topics that would draw me into cryptozoology, I found myself partaking in books on weird animals, strange beasts, and a subject called romantic zoology in the late 1950s.

Sir Richard Francis Burton, for example, seems to have used the phrase "romantic zoology" a few times in connection with sea serpent legends: "Man's natural sense of personal fear probably originated the many fanciful ideas concerning the *saevissima vipera*—it is truly said, *Timor fecit deos.* The surpassing subtlety of the brute, the female supposed to devour the male, and their young their parent, with the monstrous imaginative offshoots—dragons, fiery snakes, the great sea-serpent—all such romantic zoology seems to have originated from one and the same source."[1] "The seas are little explored (1857), and there are legends of ichthyological marvels which remind us of European romantic zoology."[2]

Here's an earlier usage by another author: "Endowed with great perfection of sight and smell, the agouti nevertheless becomes, in confinement, as 'weak, insensible, and insignificant' as the writers of romantic zoology have made out the Guinea-pig."[3] Burton also uses the phrase "romance of natural history" in a footnote discussing the African unicorn in *Abeokuta and the Camaroons Mountains.*

Not until the early days of the 1960s would the word *cryptozoology* become familiar.

When I decided to write my encyclopedic milestone book *Cryptozoology A to Z* in 1999, I was aware most authors said French

1. Sir Richard Francis Burton, *Mission to Gelele: King of Dahome*, vol. 1 (1864), 96–97.

2. Sir Richard Francis Burton, *Zanzibar: City, Island, and Coast*, vol. 1 (1872), 205.

3. William Swainson, *On the Natural History of Classification of Quadrupeds* (1835), 329.

zoologist Bernard Heuvelmans had invented or coined the term *cryptozoology*, the study of unknown animals.

Cryptozoology, which literally means "the study of hidden animals," is one of the newest life sciences, and certainly one of the most exciting.

In 1955, zoologist Heuvelmans wrote his groundbreaking book in his native French. This now classic opus is, in English, *On the Track of Unknown Animals*.

The book was soon republished in several other languages, becoming an international bestseller with over one million books in print through 1995. Supposedly, we all thought, the first published use of the word *cryptozoology* (in French) was in 1959 when a book by Lucien Blancou was dedicated to "Bernard Heuvelmans, master of cryptozoology."

But thanks to Mark Rollins, an American environmental manager and artist, it was brought to my attention that the answer to the question—Who invented the word *cryptozoology*?—is not so simple. Rollins read my Heuvelmans eulogy in December 2001 and emailed me that he remembered from Heuvelmans's book *In the Wake of the Sea Serpents* that someone else actually was responsible for *cryptozoology*. I was stunned.

Speaking of two articles on sea serpents that Ivan T. Sanderson wrote in 1947 and 1948, which served as catalysts, Heuvelmans penned this incredible sentence: "When he [Sanderson] was still a student he invented the word *cryptozoology*, or the science of hidden animals, which I was to coin later, quite unaware that he had already done so."[4]

4. Bernard Heuvelmans, *In the Wake of the Sea Serpents* (1968), 508.

Intriguingly, the 1965 French edition of this same book does not contain this paragraph at all.

Scottish-turned-American zoologist Ivan T. Sanderson's first use of *cryptozoological* did not appear until 1961, in his book *Abominable Snowmen: Legend Come to Life*. Words take time to leak into the culture.

Worthy of noting is that before the widespread employment of the word *cryptozoology* coming into use and existence, it had to develop in the 1920s and 1940s. The early concept was expressed through the phrase "romantic zoology."

Romantic zoology grew out of the Victorian era of exploration and initial contact with Indigenous peoples. Western and Eastern science were interested in categorizing the entire natural world and identifying new animals through observations, folklore, and capture. *Cryptozoology* appeared after *romantic zoology*, we have always been told. But let us look at the recent evidence for this.

Willy Ley's *The Lungfish and the Unicorn* (1941) was revised and retitled as *The Lungfish, the Dodo, and the Unicorn* (1948). Both shared the subtitle *An Excursion into Romantic Zoology*.

In 2013, in Daniel Loxton and Donald R. Prothero's book *Abominable Science*, they published an early source often forgotten in discussions of the origins of romantic zoology and cryptozoology. They are to be congratulated for bringing this information to cryptozoologists' attention. Prothero wrote this:

"As early as 1941, a reviewer of Willy Ley's *The Lungfish and the Unicorn* described it as covering not only lungfish and unicorns but an array of other marvels, zoological and crypto-zoological, from the mush rush of the Ishtar Gate to the basilisk, the tatzelwurm, the sea serpent, and the dodo."[5]

5. Daniel Loxton and Donald R. Prothero, *Abominable Science* (2013), 17.

In a footnote for this sentence, Prothero noted, "The word *cryptozoological* appears at a line break and so is hyphenated: *crypto-zoological*. It is unknown if the reviewer intended the word to have a hyphen, but he used it much the same way that writers use it today. See Ralph Thompson, review of *The Lungfish and the Unicorn*, by Willy Ley, *New York Times* (April 22, 1941), 19."[6]

Ralph Thompson (1904–1979) was an American author, teacher, and editor. He was a book critic at the *New York Times* and a contributing editor at *Time* magazine, wrote reference works and translations, and was editor of the Book-of-the-Month Club from 1951 until 1975. One of his friends was Theodore Dreiser, an early member of the Fortean Society of New York City. Ivan Sanderson was also a member of the Fortean Society, named after early twentieth-century researcher and author Charles Fort (1874–1932), who detailed cryptozoological topics in his books.

Intriguingly, the word *cryptozoological* in the Loxton and Prothero book is hyphenated as *cryptozo-ological* (sic) in their main passage when quoting Thompson, although the *New York Times* published it exactly as *crypto-zoological*. This appears to be an error on the part of the copyeditor of the Loxton-Prothero book at the Columbia University Press, an unfortunate one since this passage was about the way the word was first written.

When Ivan T. Sanderson first used the term *cryptozoological* in 1961, published by Philadelphia's Chilton, in *Abominable Snowmen: Legend Come to Life*, it was hyphenated as *crypto-zoological* in the middle of a sentence. This is significant, as this is the way the term was employed in the 1940s and 1950s. Today, as noted already, the common usage is without the hyphen.

6. Daniel Loxton and Donald R. Prothero, *Abominable Science* (2013), 17.

Willy Ley's *The Lungfish and the Unicorn* (1941) came out during the beginnings and was buried under the war news. Willy Ley left Germany in 1935 (because of the Nazis), and all his writings after 1935 were in English.

It was more widely noticed when Ley's book was republished after World War II. *Time*, on October 4, 1948, headlined "Science: The Romantic Zoologist" and starts, "Big-Game Hunter Willy Ley has returned from safari. Having tracked his prey through the dank undergrowth of large public libraries, he has put his trophies on exhibition in a newly published book, *The Lungfish, the Dodo, and the Unicorn* (Viking; $3.75)."[7]

Ley's book appeared in Germany only as a translation, although Ley had published cryptozoological material in Germany before he emigrated, for example, on living dinosaurs in *Vorwärts*, the newspaper of the Social Democrat Party.

His book *The Lungfish, the Dodo, and the Unicorn* appeared in Germany, where it was published by Kosmos (Stuttgart) in 1953 under the title *Drachen, Riesen*. But Ley does not employ any name like *Romantische Zoologie/Tierkunde* (as it would be rendered in German).

Cryptozoology grew out of romantic zoology, and Jason Offutt continues to chase monsters today in the same tradition. Thank goodness, updating the field with his fantastic detailing.

Loren Coleman, Director
International Cryptozoology Museum
October 31, 2024

7. Read the full article here: https://time.com/archive/6601634/science-the-romantic-zoologist/.

"The oldest and strongest emotion of
mankind is fear, and the oldest and strongest
kind of fear is fear of the unknown."

—H. P. Lovecraft

Introduction

Hello, dear reader.

In my 2019 work, *Chasing American Monsters*, I reported on unidentified creatures from all fifty of the United States. In *Chasing North American Monsters*, I tackle the entire continent, from beyond the Arctic Circle to Panama. The monsters in this book range from big to small, slimy to hairy. Hmm. Let's start with big and hairy.

Gimme a second first.

Cryptozoology isn't an exact science, mainly because the only exact zoological sciences are ones with monkeys ooking from cages, or cold bodies on examination tables. Some of our favorite animals—the mountain gorilla, the giant panda, the okapi, the megamouth shark—were cryptozoological creatures one hundred twenty years ago (the megamouth more like fifty) when something wonderful (and insulting) happened. Western scientists "discovered" them. Oh, sure, the Indigenous peoples knew of these beautiful creatures, but we never listen to the people who see them.

I am in no way bashing science. Science is how we know things. I'd love to have a couple of beers with famed astrophysicist Neil deGrasse Tyson and talk about football.

What I *am* bashing is neglect. The neglect of eyewitness accounts just because we think we know better.

Which brings me to big and hairy.

The Three-Toed Bigfoot

There are a couple of big, hairy, three-toed, bipedal beasties in this book. Logically, there shouldn't be. Three toes, hair, and bipedalism don't seem to go together. But read on. It might make a little bit of sense by the end.

Fifteen-year-old Doris Harrison got the fright of her life in July 1972 as she stood in the kitchen of her family home at the foot of Marzolf Hill in the tiny town of Louisiana, Missouri. When she last checked, her brothers, Terry (eight years old) and Wally (five), were chasing their dog through the trees, but sudden screams brought Doris's attention to the window. A monster about seven feet tall, covered in hair, stood on two legs at the edge of the yard as her brothers ran toward the house. The creature, its face obscured, held a dead dog under its arm.

"It wasn't a man, and it wasn't a bear," Doris (Harrison) Bliss told the *Columbia Missourian* in 2012, the fortieth anniversary of the encounter.

That was the first sighting that summer of the beast that would be dubbed Momo, the Missouri Monster. It wasn't the last.

A man said the monster chased him. School children claimed to have seen it outside their classroom window. And two women said they saw an ape-man that smelled like a skunk staring at them from the woods.

The most curious detail to any description of Momo was its feet. It only had three toes. Local man Clyde Penrod made plaster casts of the prints.

"It's not human at all," Penrod's daughter, Christina Windmiller, told the *Missourian*. "It has a big heel and three toes."

Although the sightings trickled to a stop after a couple of weeks, one imprint Momo left on the field of cryptozoology was those toes. Most Bigfoot prints have five toes (sometimes four) and, although the big hairy beast is often referred to as a North American Ape, its feet don't resemble an ape's with a thumb-like toe on the side. The prints appear human—just big.

However, the Momo prints aren't alone. There are more reports of Bigfoot tracks with three toes.

In the early 1970s, while Momo was stomping around Missouri, an identical monster appeared one state down in Miller County, Arkansas. From 1971 to 1974, the Fouke Monster, a seven-foot-tall hairy, humanlike creature, was reportedly seen in an area over hundreds of miles.

The Fouke Monster was first reported to the local newspaper when Bobby and Elizabeth Ford claimed the creature reached through a screen window of their home the night of May 2, 1971, according to the *Texarkana Gazette*. Bobby and his brother Don chased the monster away, shooting at it in the darkness.

Along with the shredded screen and scratch marks on the siding, the beast left three-toed footprints around the house. The monster, and its strange feet, terrorized the Fords for days.

Locals claimed to have seen the apelike monster often in the area. Local gas station owner Scott Keith discovered more footprints in a soybean field—these also had three toes.

Taking another step south to the commonwealth of Louisiana, locals there have talked of the Honey Island Swamp Monster since the early 1960s. Like Momo and the Fouke Monster, the Honey Island Swamp Monster is bipedal, seven feet tall, covered with hair, and smells of skunk and rotting flesh—it also has three toes.

Retired air traffic controller Harlan Ford was the first to report the monster in 1963. He also claimed to have found three-toed prints of the beast in 1974. Another local man, Ted Williams (not the baseball player), also said he saw the swamp monster, according to *Nexus Newsfeed*.

Sightings of a seven-foot-tall Bigfoot that leaves three-toed prints have also been reported in Lockridge, Iowa. In October 1975, farmer Herb Peiffer saw a "black-haired thing in the cornfield" near Lockridge, according to an article in the *Milwaukee Sentinel*. As Peiffer, and later witnesses, attested, the creature left ten-inch-long, three-toed prints. Similar prints were discovered in Minnesota in 1989 on the Vergas Trail east of Fargo, according to the *Detroit Lakes Tribune*. If this is a hairy, manlike creature with three-toed feet, what exactly is the Bigfoot that roams a corridor down the center of the United States? A corridor close to the Mississippi River?

People have speculated Bigfoot is anything from the ten-foot-tall Pleistocene Asian ape Gigantopithecus, to remnant Neanderthals, to misidentified bears, to at least one claim that it is Cain of the Bible. None of these guesses explain why some Bigfoot—which appear to be mammals—may have only three toes on each foot.

There are only three mammals that have three toes on the hind limbs. The rock hyrax (a twenty-inch-long native of Africa and the Middle East), the guinea pig (an eight-inch-long rodent native to

South America), and the sloth (a thirty-inch tree-dwelling mammal native to Latin America).

Of these, only the sloth has placed any foot in Bigfoot circles and is mentioned in this volume: the Cave Cow of the Yucatán Peninsula and the Saytoechin of the Yukon. There's also the Mapinguaris of the Brazilian rainforest.

The giant ground sloth (Megatherium), whose range stretched from South America to northern Canada during the Pleistocene Era, weighed as much as four tons and grew to twenty feet in length. It was an omnivore (as Bigfoot is purported to be) and could move on two legs as well as four. And, of course, it had three toes.

Could the giant ground sloth have survived in the unknown corners of the Americas for the past eleven thousand years? According to Native legends, yes.

The Mapinguaris is a creature villagers in the Brazilian rainforest claim to have encountered as recently as the 1990s, according to the *Los Angeles Times*. It is around seven feet tall, covered in reddish-to-black fur, stands on its hind legs, is omnivorous, and smells like rotting flesh. Ornithologist David Oren has gone as far as proposing the Mapinguaris may be a living giant ground sloth.

A similar creature lives in the legends of the Inuit peoples of Canada's Yukon territory. The Saytoechin is a creature larger than a grizzly bear that feeds on tree bark and beavers.

An Inuit family fishing at Tatchun Lake in central Yukon in the mid-1980s witnessed an animal "eight or nine feet high, bigger than a grizzly bear…coming toward them," per the British Columbia Scientific Cryptozoology Club. When the family was shown pictures to identify the creature, they pointed to drawings of a giant ground sloth.

Could this creature, thought dead for eleven thousand years, be responsible for three-toed Bigfoot reports through the center of the United States? Maybe. In the field of cryptozoology, you never know what's out there.

Welcome to the land of monsters. Enjoy.

Chapter 1
Belize

Sea, rainforest, mountains, Maya ruins, a barrier reef, and world-class food; the second-smallest country in North America has a lot to offer its on-average 370,000 yearly visitors. The Belize Barrier Reef, the second-largest barrier reef in the world, offers excellent boating and scuba diving. Especially interesting is Belize's Great Blue Hole, a 1,043-foot circle that dives 406 feet into the sparkling Caribbean waters. The hole is home to sponges, corals, sea turtles, and more sharks than a Hollywood blockbuster, including hammerheads, nurse sharks, bull sharks, and Caribbean reef sharks. Belize doesn't neglect the land; more than 75 percent of the country's rainforest is protected by the government. However, the monsters aren't protected, because they can usually protect themselves.

Sismite

Bigfoot-like creatures have been reported all over the world; from the Pacific Northwest to Europe, Asia, Australia, and South America. Belize is no exception. The Sismite is a hairy, gorilla-like beast with a human head that lives in the forests and jungles of the country. Not friendly with man, the Sismite have developed a taste for human flesh, especially male flesh.

The creature has a dislike for men. If a man looks into the eyes of a Sismite, he will die within a month, but if a woman looks the monster in the eyes, she will live a longer life. But the Sismite isn't kind to women; they kidnap women to use for breeding. These creatures live in caves high in the mountain forests, far away from human habitation, and are rarely spotted.

Curiously, this Central American Bigfoot—with brown-to-black fur—has backward-facing feet and no thumbs. It is terrified of dogs and has a distaste for water. One sure method of getting rid of a Sismite is to set its hair on fire.

The Sismite is prone to attack people who wander into the forests during Christian religious holidays and Sundays. Also a protector, this monster has been known to kill anyone involved in harming the forest.

Cadejo

Legends of spectral dogs exist worldwide. Black dogs haunt the British Isles, Germany, New England, and the southernmost parts of North America. Called the Black Shuck in England and the Black Dog of Hanging Hills in Connecticut, the evil ghost dog from Mexico through Panama is known as the Cadejo.

In most stories, the Cadejo appears as a black dog with blue eyes that turn red when it is angered. It travels through lonely areas

CADEJO

at night, attacking those who are foolish enough to wander in the dark. This black dog is said to have been created by the Devil, and its opposite, a white Cadejo, by God. The white Cadejo protects those who fall victim of the black Cadejo.

However, the Belize story is a bit different. In Belize, beware the white dog.

The white dog is the demon dog, choosing the color white to trick those who would be afraid of a black dog. The black Cadejo is holy and is known to escort wanderers and drunken men who walk by night.

Either dog, however, can change their appearance at will, sometimes becoming the size of a horse, or others possessing the hooves of a goat. In any form, the sound of dragging chains accompany the Cadejo. The Spanish word for "chain," *cadena*, is thought to be the source of the name *Cadejo*.

The best way to avoid the demon dog? Stay home when the sun goes down.

El Duende

If black dog legends are common across the world, legends of little people are ubiquitous. Europe has its elves and trolls, New England has its Pukwudgies, Hawaii has its Menehune, Indonesian islands have their Ebu Gogo, and Belize has *el Duende*—"the goblin."

This mischievous entity is described as a dark little man (two to three feet tall) dressed in a big floppy hat and a fancy green suit who lives in either a cave in the forest or in the walls of a child's bedroom. Children seem to be the goblin's favorite prey, tricking them into walking deep into the forest, or simply creeping out from the walls and cutting off a toe.

Of course, much like the legends of stories of other similar entities across the world, el Duende can sometimes be helpful, leading lost children, hikers, and adventurers safely from the woods.

Tata Duende, or Grandfather Goblin, is similar in appearance and behavior to el Duende, although he appears as an ugly old man dressed in animal hides and carries a stick. Much like Belize's Sismite and the Jumbee, la Ciguapa, and Douen of Caribbean folklore, Tata Duende's feet are backward. Also, like the Sismite, Tata Duende has no thumbs.

This dwarf is seen at night, proceeded by a distant whistle (it's a trick! That means he is nearby). Although Tata Duende is also mischievous, he is a protector of the forest and will help lost people and cure them if they are injured. Just don't dig holes or fell trees in his forest. He'll lop off your head.

X'tabay

Men who wander the forests and hills of Belize alone not only need to worry about the backward-footed Sismite or el Duende; they have to worry about a woman with a comb.

The woman stands beneath the Ceiba pentandra—the sacred tree of the Maya. Her hair is long, black, and silky, and frames a face more beautiful than any a man has ever seen. She combs that hair and bats her eyes until the man stops to talk with her. They soon make love beneath the tree, after which the woman kills her lover. In some stories, she rips into the man's chest and pulls out his beating heart. In others, she transforms into an enormous snake and swallows him.

According to legend, this shape-shifting demon was once a promiscuous Maya woman who was jealous of the attention her pure sister received from the people of her village. Upon death, wicked spirits gave her the ability to return to Earth, and she spends her time killing lust-filled men, the type who violated her in life.

Chapter 2
Canada

Oh, Canada.

Known as the Great White North, Lumberjack Country, the Land of Maple, or just the North, this country of 38.25 million people (and a million moose) stretches 3.85 million square miles (or 9.98 million square kilometers, eh?) across the top of North America. The country is known for ice hockey, maple syrup, poutine, some of the most polite people in the world, and majestic scenery (that includes the Rocky Mountains, arctic tundra, forests, prairies, and the world's longest coastline at 243,042 kilometers). Ten provinces and three territories cross this British commonwealth from the Atlantic to the Pacific, but, despite the enormity of the country, around 90 percent of the population lives within one hundred fifty miles of the border to the United States. It gets cold up north. Canada produces oil, beef and poultry, wheat, wood, manufacturing (trucks and aircraft parts), iron, and steel. Famous people from Canada include actors like Michael J. Fox,

William Shatner, Keanu Reeves, and Ryan Reynolds; musicians such as the band Rush, and Avril Lavigne; and an uncomfortable amount of professional wrestlers.

It's also known for monkeys with really, really big feet.

Alberta

Alberta is one of ten provinces and three territories that make up the nation of Canada. It is the sixth-largest province or territory at 255,541 square miles (slightly smaller than Texas), the fourth most populous with a little more than four million people, and boasts a diverse geography. Alberta contains prairie, desert, forest, more than six hundred lakes, glaciers, and mountains, including the Canadian Rockies. The province produces 59.6 percent of Canada's feeder cattle and averages $6.1 billion in wood production. You want monsters? Alberta's got them—hairy and scaly.

Bigfoot

People have seen Bigfoot throughout Canada and in forty-nine of the fifty United States.

The earliest report in Alberta is from January 7, 1811, when David Thompson, a Northwest Company surveyor, saw something in the Rocky Mountains he didn't understand. He wrote about it in his journal (spelling and grammar not corrected).

"I saw the track of a large Animal—has 4 large Toes abt 3 or 4 In long & a small nail at the end of each. The Bal of his foot sank abt 3 In deeper than his Toes—the hinder part of his foot did not mark well. The whole is about 14 In long by 8 In wide & very much resembles a large Bear's Track. It was in the Rivulet in about 6 In snow."

Thompson addressed the prints decades later in his memoirs. "We were in no humour to follow him; the Men and Indians would have it to be a young mammouth and I held it to be the track of a large old grizzly bear; yet the shortness of the nails, the ball of the foot, and its great size was not that of a Bear, otherwise that of a very large old Bear, his claws worn away, the Indians would not allow."

Reports of the enormous, hairy, apelike human have continued across the decades. According to the Alberta Sasquatch Organization website, there have been at least eighty-seven Bigfoot reports in the province since the 1940s. One Alberta man is so serious about the existence of the creature, he's sued British Columbia.

Todd Standing of Edmonton, director of the documentary *Discovering Bigfoot* (2017), has established a group that filed a lawsuit against the Canadian province to recognize the reality of the creature.

"We're going in with PhDs, with wilderness experts beyond myself, with wildlife biologists, with fingerprint experts. We're going to prove so beyond a reasonable doubt that this species exists," Standing told Canada's *Global News*. "When we prove that and we're successful, the species will be recognized as an Indigenous wildlife species and then fish and wildlife—in California, in Canada, in the United States, everywhere."

The group claims to possess Sasquatch DNA samples that will definitively prove it exists.

"Half of my evidence is from Alberta," Standing told *Global News*. "I've had lots of success in Alberta."

Unfortunately for Standing, according to CBC News, the hair sample he collected from a tree in British Columbia in 2014 turned out to be human.

British Columbia's Supreme Court dismissed the lawsuit on August 31, 2018, when Justice Kenneth Ball said the man had "no legal standing to bring such a claim" as the existence of Bigfoot.

Kinosoo

With a surface area of 144 square miles and a depth of 325 feet, Cold Lake is one of the largest and deepest lakes in Alberta and is split by the Alberta and Saskatchewan border. The lake features recreational sports, including fishing and boating. The early Chipewyan Indians called the lake Big Fish Lake, although it was the Cree Indian title of Coldwater Lake that gave the body its name.

It was here the local First Nations tribes told white settlers the story of the great fish Kinosoo.

According to local legend, a young First Nations man often paddled a canoe near the shores of the lake to visit a young woman he hoped to marry. Although sticking near the shores was a safer way to travel over the often turbulent lake, one night when the winds were low, the young man decided to cut across French Bay when he was attacked. An enormous fish rushed at his boat and bit it in two, pulling the man into the water—all while the woman he hoped to marry watched helplessly from the far shore. The only things that remained were a paddle and pieces of the broken canoe.

Wendigo

Fort Kent, a tiny town twelve miles south of Cold Lake has its own story of a monster. Tales of the Fort Kent Wendigo begin in the 1870s when farm animals begin to disappear from around Fort Kent—and so did the farmers.

WENDIGO

According to First Nations legends across North America, the Wendigo is a creature that was once a man who succumbed to the whims of an evil spirit and resorted to cannibalism. A Wendigo appears as a tall, furry, emaciated creature with bones pushing against its skin. The beast has glowing red eyes and the head of a deer.

The story of the Fort Kent Wendigo began when the Cree Indian trapper Swift Runner returned alone to the area from his winter camp, although his family had gone to camp with him. Swift Runner explained his wife killed herself and his children starved to death. When North-West Mounted Police forced Swift Runner to lead them to his camp, they found the remains of the trapper's family with evidence they had been butchered. Swift Runner admitted to eating his family, but blamed the horror on the spirit of a Wendigo. He was tried for the murders and sentenced to hang. Swift Runner was executed at Fort Saskatchewan, Alberta, on December 20, 1879.

Whether the spirit of a Wendigo possessed Swift Runner or he had simply succumbed to ravenous hunger during the winter is uncertain. However, if it was a Wendigo, the spirit didn't leave the tiny town of Fort Kent.

Smallpox hit the town in 1921, decimating the already small population and overwhelming the town doctor Thomas Burton. When Burton's wife died during the outbreak, he ate her. Then he ate more people. Once he was out of readily dead people, he began killing the town residents for food. When the few remaining townsfolk confronted the doctor, he fled into the forest.

Dogman

Legends of bipedal dogs stretch across North America. Alberta is no exception. Researcher Rona L. Anderson told the *Edmonton*

Examiner of a reported two-legged dog running across the road in front of a car.

"It had the shoulders of a man and the head and face of a dog, with very, very short hair, which was almost a gray color," Anderson told the *Examiner.*

The monster, or something similar, has been seen in the area before. It's described as standing six feet tall with short fur that ends in a long, bushy tail. Besides the fact it stands on two legs, the dogman's most frightening characteristic is its human hands.

Hyena

Stories of hyena-like creatures are not as uncommon as they should be. This wouldn't be surprising if we were in the Pleistocene Epoch when the giant hyena Chasmaporthetes roamed the land. But the time is now, and Chasmaporthetes has been extinct for about 780,000 years. The descriptions of the big doglike creatures also suggest the dire wolf, which died out about 9,440 years ago, or the much longer extinct Hyaenodon (died out 23 million years ago) and the hyena-like dog Borophaginae (died out 2.5 million years ago). But whatever the improbably historic possibilities of the cryptid canines sighted in Canada and the United States, the name it goes by in Ioway Indian folklore is Shunka Warak'in (carries off dogs).

Although the Shunka Warak'in is best known for the creature shot by rancher Israel Ammon Hutchins in 1896 in Montana, it has also been seen in Illinois, Nebraska, and Alberta. The most notable Alberta sighting was near the small town of Legal, just north of Edmonton, in 1991. Several people saw a creature that resembled a hyena pacing near the entrance to a park.

Banff Merman

Alberta's Lake Minnewanka lies within the Banff National Park. At around seventeen miles long, it is the second-longest lake in the parks that encompass the Canadian Rockies; its maximum depth is 466 feet. Lake Minnewanka also has a long history of...mermen?

Yeah, mermen. Just not in the lake. Not at all. Seriously. That's the last time I'll talk of Lake Minnewanka.

As the story goes, the First Nations peoples who lived near Lake Superior (not in Alberta) told stories of water spirits that inhabited the lake. One of those water spirits was a merman. The creature—a man-fish about the size of a seven-year-old child—would surface near anyone swimming in the lake, terrifying them with its ugliness. So, what happened when voyageur Venant St. Germain saw the creature on the shores of Lake Superior in 1782? He killed it.

Sometime over the next one hundred twenty years, the merman made its way one thousand-plus miles from Ontario to Alberta.

The merman is in the Trading Post in Banff, a town of nearly eight thousand people. The Trading Post was started in 1903 by Norman Luxton and was maintained by the Luxton family until 1961. Norman Luxton was a newspaperman who worked in Winnipeg, Calgary, and Vancouver. In Vancouver in 1901, he met an adventurer named John Voss, and, while drunk, the men decided they would buy a dugout canoe and try to sail around the world. They made the attempt, but Luxton fell out of the boat somewhere in the South Pacific and suffered coral poisoning. He was finished with the trip.

When he arrived back in Canada, a doctor suggested Luxton visit Banff for its supposed healing waters, and once there, Luxton decided to stay. He bought the town's newspaper, built the Trading Post, a hotel, and a local museum...but it's the merman lying on its stomach in a display case inside the Trading Post everyone talks about.

And if the Banff merman was ugly in 1782, it's only gotten uglier. The lower part of the mummified creature is, as advertised, a fish. The emaciated upper half is all ribs, its grotesque head topped with gray hair. Its appearance is extremely similar to P.T. Barnum's famous Fiji Mermaid, which is composed of the upper half of a blue monkey and the lower half of a salmon.

Speculation is, Luxton bought the merman, published a story about voyageur Venant St. Germain killing the beast in Canada, and put the creation in his store to attract customers.

Sounds about right.

British Columbia

British Columbia, the westernmost of Canada's provinces, is the only province that touches the Pacific Ocean. Of its 364,764 square miles of total area, 231,661 square miles are forestland, the combined size of Germany and France. Of that woodland, most of it is old-growth forest—the same forest that graced British Columbia before Europeans settled the area. The province has sixteen mountain ranges, including the Canadian Rockies, seven national parks, and two of the deepest lakes in North America: Quesnel Lake (1,660 feet) and Adams Lake (1,499 feet). All this nature makes British Columbia a haven for tourism. It's also the perfect home for monsters.

Ogopogo

Canada's most famous lake monster Ogopogo is supposed to live in Okanagan Lake, fifty-two-foot-long, two-mile-wide body of water next to Kelowna, the third-largest city in British Columbia (population 194,882). The lake, a little more than a two-hour drive north from Washington State, is 761 feet deep, and has plenty of room to hide a mythical beast.

First Nations legends describe Ogopogo as a fifty-foot-long serpent that to Western science most resembles a Basilosaurus, an extinct sea mammal (the Basilosaurus has *saurus* in its name due to an early misidentification as a sea reptile). Although known by the local First Nations people for hundreds of years, the name *Ogopogo* is of Western origin. According to the *Penticton Herald*, the Okanagan First Nations people call the creature N'ha-a-tik (sacred of the water).

Reports of sightings by white settlers began in 1872. Although no human was ever attacked by Ogopogo, a team of horses was dragged underwater by a mysterious creature in the 1880s. According to an article in the *Los Angeles Times*, officials in the British Columbia government were so concerned about the creature that in 1926 they discussed arming the ferries that hauled people and freight across Okanagan Lake.

Modern sightings include a June 20, 1986, fishing expedition when Lionel Edmond saw what he described to the *Los Angeles Times* as enormous.

"I turned around and I saw this dark thing coming through the water," Edmond said. "It looked like a submarine surfacing, coming up toward my boat... We could see six humps out of the water, each hump about ten inches out of the water and each one creating a wake... We saw no head or tail, but it must have been some fifty or sixty feet long."

Some sightings since 2005 have included mobile telephone video, but reveal nothing more than blurry dark shapes in the water.

Cameron Lake Monster

Ogopogo isn't Canada's only lake monster. Heck, it's not even British Columbia's only lake monster. But if the Cameron Lake Monster exists, it's not the only legend associated with this cold 87.4-acre body of water.

According to stories of this 141-foot-deep lake on Vancouver Island, an airplane crashed into the water in 1968 and wasn't recovered until 1983. Unverified stories claim the bodies of the pilot and passenger were recovered in perfect condition due to the frigidness of the water. A train was supposed to have plummeted into the lake early in the twentieth century, and a tractor-trailer that flipped off the highway and sank into the lake in 2016. The driver and passenger made it to safety.

But it's the monster Cammie that makes the headlines.

In 2009, a group of researchers probed the waters of Cameron Lake in response to numerous sightings of something large and unknown swimming beneath the waters of the lake. According to *Live Science*, the British Columbia Scientific Cryptozoology Club has investigated the lake since it began receiving reports of the creature in 2004. "Witnesses have been describing what looks like a dark creature in the lake," the club's cofounder John Kirk told *Live Science*. Witnesses have described Cammie as long, black, and serpentine.

"Some of the earliest sightings go back to the 1980s," another club member, Adam McGirr, said when interviewed on the breakfast television talk show *Canada AM*. "Most recently, though, there seems to have been an increase in the number of sightings."

One of those sightings was by Brigette Horvath, who was driving by the lake in 2007 when she saw a large beast swimming in circles just beneath the surface, according to *CTV*. Club members speculated Cammie could be anything from a sturgeon to an eel to an enormous trout. The group detected something large swimming in the lake in 2009, but stopped the investigation when the underwater camera it was using became damaged.

Cadborosaurus

This sea serpent with a horselike head, large front-facing eyes, and flippers on a body made of coils has been reported off the coast of British Columbia for hundreds of years. First Nations history is filled with stories of Caddy, and the creature has been carved into petroglyphs all along the Pacific coastline as far north as Alaska and as far south as San Francisco Bay. Sightings by European settlers date back two hundred years.

The crew of the fur trading ship *Columbia* saw a sea serpent off the coast of British Columbia in 1791. In 1897, a man reported seeing a twenty-four-foot, long-necked animal swimming near Queen Charlotte Islands. A brown sea creature with a nine-foot-long neck was reported in Johnstone Straight in the early 1900s. Similar reports have continued until this day.

Various carcasses purported to be Caddy have been retrieved in British Columbia, such as the twelve-foot-long remains cut from the stomach of a sperm whale in Naden Harbour in 1937, forty-six-foot-long remains discovered at Vernon Bay on Vancouver Island in December 1947, and in 1962 a twelve-foot-long body with a head like an elephant was discovered near Ucluelet. Although a scientific explanation was offered for each of these cases, nothing was conclusive.

Fisherman William Hagelund claimed he caught a live sixteen-inch-long infant Cadborosaurus off De Courcy Island in 1967. The creature had scales, two flippers, and a long, flat tail. He placed

CADBOROSAURUS

it in a bucket of water to have it examined by scientists later, but for some reason decided to put it back in the water instead.

Devil Monkeys

Devil Monkeys are meter-tall bipedal, vicious, baboon-like creatures with hind legs like a kangaroo, three toes on each foot, and a long bushy tail. Devil Monkeys have not only been known to attack livestock and pets, they also seem to attack humans with no provocation.

Although these creatures are reported across the Appalachian Mountain region of the United States, numerous reports have come from British Columbia. One of the most famous accounts was by Gordon Ferrier who saw a meter-tall hairy biped with a bushy tail and three toes near the Mamquam River in June 1969. This encounter was investigated by famed Canadian Bigfoot researcher John Green.

Albert Ostman's Abductor

Although Bigfoot-like creatures are reported all over the world in a variety of topographies, with its great expanses of forests and mountains, British Columbia could be the textbook Sasquatch habitat. With eleven people per square mile, this gives the big hairy fellow a lot of places to get away from those pesky humans. That is, if it wants to. British Columbia is home to one of the most famous Sasquatch encounters in North America.

In 1924, Canadian lumberjack and prospector Albert Ostman slept at his campsite near Toba Inlet when he claimed a great bipedal beast lifted him in his sleeping bag and tossed him over its shoulder. The creature carried Ostman for what he believed to be three hours before it stopped on a plateau and put him down. What Ostman saw shocked him. He was surrounded by four huge, hairy, humanlike creatures; three adults; and a juvenile. The tallest was at least eight feet tall.

The creatures kept him for six days. Although the Sasquatch family (which is how he viewed them) didn't harm him, he thought the biggest beast may have taken him to become a mate for the juvenile female. On the sixth day, the large male who had kidnapped him became interested in Ostman's snuff. When Ostman gave the creature the tobacco, it flew into a sneezing fit and Ostman escaped.

He held on to his encounter until 1957, when other people started coming forth with their own stories.

Thetis Lake Monster

Thetis Lake, Canada's first regional conservation area, is a picturesque forested park on Vancouver Island, popular for fishing, boating, and hiking. In 1972, this peaceful recreational area was terrorized by a bipedal monster.

Two teens reported to the Royal Canadian Mounted Police that a five-foot-tall lizard man had attacked them. The creature had spines on its skull; its hands and feet had three webbed digits apiece. The monster's attack cut one boy's hand before they escaped. Four days later, *Victoria Daily Times* reported two men saw a scaly, silvery-blue humanlike creature across the lake.

Although various explanations were given for the lizard man (including an escaped pet tegu lizard that can grow to more than three feet in length), nothing terrestrial quite matched the lizard man's description. It did, however, resemble South Carolina's Lizard Man of Scape Ore Swamp and Ohio's Loveland Frogman (both featured in *Chasing American Monsters*). One explanation for the Thetis Lake Monster stuck—one of the original witnesses said his friend who turned in the report was simply a big liar.

Manitoba

Manitoba has a special place in the world of monsters. The word *cryptid* was first used by Manitoba's John E. Wall in 1983 to describe animals sought by cryptozoologists. The province of Manitoba stretches from the US border of North Dakota in the south to the territory of Nunavut to the north. The land of Manitoba consists of prairies, farmland, mountains, lakes, rivers, and tundra. With 249 miles of Hudson Bay coastline to the northeast, Manitoba is the only Prairie Province (which includes Saskatchewan and Alberta) to have a saltwater border. Manitoba has more than 110,000 lakes that fill 15.6 percent of the province's surface area. Maybe that's why Manitoba has so many lake monsters.

Manipogo

Lake Manitoba, in the central portion of the province, is the world's thirty-third-largest freshwater lake, although it's only the third-largest lake in Manitoba. It is 125 miles long by 28 miles wide with an area of more than a million acres. That's a lot of water. However, it is only twenty-three feet deep at its deepest. However, this lack of depth doesn't mean the lake might not be hiding a brown fifty-foot-long monster.

Although First Nations stories of the creature go back centuries, white settlers began seeing it in the 1800s. The first well-documented sighting of Manipogo (named in 1957 after Canada's most legendary lake monster Ogopogo) seems to be from 1909. Hudson's Bay Company fur trader Valentine McKay claimed to see a huge creature in the water in September of that year. Twenty-six years later in 1935, timber inspector C. F. Ross and a friend reported encountering a monster there that looked like a dinosaur.

Sightings continued over the decades. In 1948, a man saw a creature that stretched its neck nearly seven feet out of the water and roared like a dinosaur. In 1989, a family from Minneapolis saw several humps rise from the water. In 2004, commercial fisherman Keith Haden said his nets were destroyed by what appeared to be enormous bites.

However, none of these sightings were like one from the 1960s. This time, someone got a picture. Two fishermen, Richard Vincent from North Dakota and John Konefall from Dauphin, Manitoba, were returning to their camp on August 12, 1962, when they saw something odd about one-hundred-eighty feet from their boat and took a photograph, per a 2013 article in the *Winnipeg Sun*. It is the only photograph of Manipogo known to exist.

Winnipogo

Lake Winnipegosis, the second-largest lake in Manitoba at 121 miles long and 32 miles wide with an area of more than 1,300,000 acres, is quite a bit deeper than its neighbor Lake Manitoba at sixty feet. This recreation lake is also the home of another supposed lake beastie.

Winnipogo (like Manipogo is named after Ogopogo) is a brown, twenty-five-foot-long serpent with a horn on its head. Although First Nations people have legends of the beast that stretch back centuries, one of the first cases documented in a newspaper was in a 1901 edition of the *Dauphin Press*. Sightings were also reported in 1909 and 1935. In the 1930s, Oscar Frederickson discovered a bone he couldn't identify on the shores of the lake. Dr. James McLeod of the University of Manitoba said the bone could be a vertebra of an extinct whale.

WINNIPOGO

Modern sightings of the beast are scant, which led the Winnipogo to be named the Most Threatened Lake Monster of 2013.

If there is a monster in Lake Winnipegosis, it may not be Winnipogo; Lake Winnipegosis and Lake Manitoba are connected. Any serpent seen in Lake Winnipegosis may just be Manipogo on vacation.

Sasq'ets

Sasquatch, most probably taken from the First Nations Salish tribe's word *Sasq'ets* meaning "wild man," have been seen in Canada since there have been people in Canada to see them. Given Manitoba's 101,545 square miles of forestland, there have been plenty of reports of this enormous apelike creature, like this one from April 21, 2005, in the Norway House Cree Nation.

When ferry operator Bobby Clarke took a routine trip across the northern tip of Lake Winnipeg, he didn't know he would see something that would make headlines around Canada, according to the CBC. A Sasquatch stood on the bank. Clarke had a camcorder on the ferry and captured two minutes and forty-nine seconds of the creature. The video shows the beast to be about three meters tall.

"It's not a bear or human walking around," Clarke's father-in-law John Henry told CBC News.

Hybrid Bears

Bears seem to more heavily populate children's stories than nature, but they are dangerous; there were ninety-two attacks in Canada and the United States from 2010 to 2013 alone. And the two most dangerous bears, grizzlies and polar, have been mating to create a Super Bear.

Super Bear is not an official name, although it should be; instead, they're called Pizzlies, or Grolar Bears.

Grizzly and polar bears almost never meet in the wild, and crossbreeding is nearly out of the question. Grizzlies are coming out of hibernation in the forest during the polar bear mating season, which occurs on the ice. However, per Canada's *Global News*, climate change has seen these unlikely mates growing more common. Nine DNA-identified hybrids have been confirmed since 2006. These bears generally have a head that more resembles a polar bear with huge grizzly shoulders. Their fur ranges from tan to off-white.

Manitoba may be the future home of these hybrid bears. Grizzly sightings have increased in Wapusk National Park near Churchill, Manitoba, which could soon overwhelm the polar bear population. Churchill, dubbed the Polar Bear Capital of the World, may someday be the Pizzly Capital of the World.

New Brunswick

New Brunswick, one of the smallest Canadian provinces at 28,150 square miles, is nestled on Canada's eastern shores between Nova Scotia and Quebec, sitting atop the American state Maine like a pompadour. First Nations peoples have lived in the area since at least 7000 BCE. New Brunswick was also part of Vinland, a section of North America explored by the Vikings around 1000 CE. The province is covered by fifteen million acres of forestland, more than sixty rivers, and boasts the Bay of Fundy where tourists flock to watch whales. New Brunswick's Saint John was the first incorporated city in Canada. It's also home to unexpected creatures, like the...

Mountain Lion

Mountain lions exist. We know this; we see them in zoos, on nature programs, and sometimes—at least in western North America—up close and personal (I stumbled upon tracks in Colorado once after a rainstorm. However big you think a mountain lion is, according to those tracks, it's bigger than you think). European farmers who immigrated to North America killed the Eastern Mountain Lion (also called the Eastern Cougar or Eastern Panther) to protect their livestock, much like the Australians did to the Thylacine. By the late 1800s, the Eastern Mountain Lion was considered endangered. A specimen was last seen in New Brunswick in 1932 and in Maine in 1938. By 1940, this subspecies was gone, although it wasn't declared extinct until 2015.

But is it extinct?

Naturalist Bruce Wright was bigger than life. A lifelong outdoorsman, Wright became a forester through the University of New Brunswick, working as a forest biologist after graduation. During World War II, he came up with the idea for frogmen who would use scuba gear to infiltrate enemy territory. During the war, he used his disposition in Myanmar (then Burma) to study marine life. After the war, he studied black ducks in the Canadian Maritimes (eastern provinces), but one of his personal interests was extinct species he believed could still be alive. One he was particularly interested in was the Eastern Mountain Lion.

During Wright's quest for the endangered (and assumed extinct) animal led him to collect hundreds of sightings and indeed discovered what is widely considered as the last Eastern Mountain Lion. When that cat died, the entire species was considered dead as well.

Since then, there have still been reports of mountain lions in New Brunswick, although they are credited to a species of mountain lion from the western portions of North America that have wandered far afield. But are they?

Lake Utopia Monster

Lake Utopia, in Charlotte County near the Maine border, is a four-and-a-half-mile-long, two-mile-wide body of water that reaches an average depth of around thirty-six feet. A popular destination for recreation, Utopia Lake is also known for the Lake Utopia Lake Monster, Old Ned.

Stories of the creature began before the arrival of European settlers when the local Maliseet First Nations tribes claimed a large underwater creature that looked somewhat like a whale would chase people canoeing on the lake. European settlers began reporting encounters with the monster in the 1800s, and they continue to this day, usually every three to five years. Although the whale explanation is a stretch, it could be possible. Lake Utopia is connected to the Bay of Fundy via the Magaguadavic River, although it wouldn't exactly be smooth swimming for a whale.

One of the most famous sightings was in 1867 when sawmill workers saw a thirty-foot-long, ten-foot-wide creature splashing in the lake. Similar sightings occurred a year later in 1868, again in 1872, and 1891. Modern sightings include a 1996 report from a couple, Roger and Lois Wilcox, who saw a fifty-foot-long creature swimming in the lake. It swam up and down like a mammal, not side to side like a fish or reptile.

TOTE-ROAD SHAGAMAW

The Tote-Road Shagamaw

From the early days of the European settlement of New Brunswick, workers in lumber camps began to report a creature with the front paws of a bear and the back legs of a moose—the Shagamaw. To confuse anyone who followed it, the Shagamaw would switch off walking on its hind legs to its forelegs. One set of tracks would extend on tote roads (trails used to carry supplies to a lumber camp) only about 440 paces because that's as high as the Shagamaw could count. Then it would switch to the other legs and go 440 more paces before switching back.

According to the book *Fearsome Creatures of the Lumberwoods* by William T. Cox (1910), the Tote-Road Shagamaw was a horrifying creature to see, but it was shy and harmless. The Shagamaw also had a peculiar diet; it ate discarded boots, mittens, and anything else lumbermen dropped along these desolate roads.

Laverna Wood Monster

Canada wouldn't be Canada without Bigfoot reports. There have been more than seven Bigfoot sightings in New Brunswick during the past thirty-five years.

According to the Bigfoot Field Research Organization, more than twenty Canadian and US soldiers, based in the Fifth Canadian Division Support Base Gagetown, were training near Laverna Wood in 1990 when unidentified noises from the trees stopped them in their tracks. The soldiers heard an enormous creature thunder through the brush and could feel its impact when its feet hit the ground. When it vocalized, it sounded like a baby's cry amplified "500 times louder."

Although the creature remained hidden by the forest, whatever made that noise was nothing the soldiers were familiar with. One soldier said, "It wasn't a bear."

Another sighting occurred in 2008 when two couples from Anfield and Saint John, New Brunswick, saw a "pitch-black, approximately eight-and-a-half-foot Sasquatch" at Skiff Lake, according to an article in the *Woodstock Bugle-Observer*. "I know a bear can stand on its hind legs and move around," one of the men told the newspaper, "but a bear can't walk on two legs the way this humanlike form (did)."

Newfoundland and Labrador

Newfoundland and Labrador, which form the easternmost province of Canada, is the land where the Viking Leif Erikson first came to the shores of North America. Newfoundland, an island, and Labrador, the mainland, are about 156,371 square miles, which makes the province one of Canada's smallest (not Prince Edward Island small, but compared to the massive territory of Nunavut, it's tiny). The population of the province is not large, at 526,000 people, which is about the size of Tucson, Arizona. The province is made up of coastline, forestland, World Heritage rock formations, and mountains; the Long Range Mountains are the northernmost arm of the Appalachians. Then there's the Nennorluk.

Labrador Nennorluk

For centuries, an enormous sea monster capable of treading on land terrorized Inuit tribes from Labrador to Greenland—the Nennorluk. This creature, whose name roughly translates to "evil polar bear," was first seen by Europeans in the 1700s. David Crantz's *History of Greenland* (1773) describes the Nennorluk as huge, with ears "large enough for the covering of a capacious tent." Inuits claimed the white creature was as large as "a huge iceberg." The monster's diet was largely seals, which the Nennorluk would completely devour, but it didn't shy away from eating humans if they happened to be in the way when it was hungry.

The Inuits believed the Nennorluk to be even larger than Crantz described. According to legend, the Nennorluk does not swim; it walks on the bottom of the ocean and is so large it can often be seen on the surface.

One-Legged Natives

When the Vikings came to Vinland (the area of Newfoundland where Leif Erikson landed in 1000 CE), they encountered something they didn't expect—native tribes of einfæting, or one-legged people.

When Erik the Red's son Thorvald Eriksson spied an einfæting one day, the meeting was deadly, according to the book *Wonderful Strange: Ghosts, Fairies, and Fabulous Beasties* by Dale Jarvis (2005).

"One morning Karlsefni's people beheld as it were a glittering speck above the open space in front of them, and they shouted at it. It stirred itself, and it was a being of the race of men that have only one foot, and he came down quickly to where they lay," Jarvis wrote. The einfæting shot an arrow that pierced Thorvald's body, killing him.

The einfæting ran and Thorvald's men gave chase, but the one-legged native escaped.

Cressie

First Nations people who lived near Crescent Lake in Newfoundland had legends of the Swimming Demon, a giant eellike creature that could appear in human form and would seduce people to follow it into the depths where it would devour them.

The first European settlers who saw something unexplained in the lake were fishermen who reported what at first appeared to be an overturned boat. When they coasted toward it to right the craft, the monster flipped over and dove beneath the waves. Other notable sightings occurred in the 1980s when scuba divers encountered a school of enormous eels, and in 2003 when people saw something at least sixteen feet long swimming on the surface of the lake.

Fairies

When people immigrate to an area, they don't just bring their personal belongings—they bring their language, their customs, and their legends. Early settlers from the British Isles found their fairy stories didn't stay home in Britain. They followed them to the New World.

European fairies aren't Disney fairies. They are most often described as gnomes, although they can appear smaller, larger, in animal shape or as glowing lights. They're also mischievous. According to the book *Strange Terrain: The Fairy World in Newfoundland* by Dr. Barbara Rieti, encounters with fairies can turn deadly. "They play tricks and lead you over the edge or a cliff," Rieti wrote. "They'll change people. Or you'll get a fairy blast when they hit you, and then nasty stuff comes out of the wound, like sticks, balls of wool and fish bones."

These fairies, like their European counterparts, have also been known to swap their sickly babies (known as changelings) for a healthy human baby, tie together the tails of livestock if they feel slighted in some way, or spirit people into the fairy realm.

In an article written by Burton K. Janes in the *(Carbonear) Compass* newspaper, back in the early days, a Newfoundland and Labrador woman walked into the hills to find a missing cow and vanished. Despite extensive searches by the townspeople, after fourteen days the woman was still missing—until a person in town had a dream of where the missing woman was. A search party went to this spot and found the woman who claimed she was "taken astray by the fairies."

Sea Monster of Bonavista

The waters that slap the coastline of Bonavista, Newfoundland, may be home to a monster. In 2000, Bonavista resident Bob Crewe drove along Lane Cove Road near Dungeon Provincial Park when he saw something he couldn't explain. "I saw its body in the water measuring about nine metres across, just lying there and moving slightly," Crewe told the UK's *Globe*. "It looked something like a rock in the water, but I knew there was no rock there."

Wanting to determine if what he saw was alive, he honked the vehicle's horn and discovered that yes, it was. The beast pushed its head out of the water. It rested atop a slender neck about five feet long. Crewe told the newspaper the neck looked like a gigantic snake. Startled by the horn, the creature took off swimming. "It seemed like it was using its body to push itself along and it was going very fast," Crewe said.

A similar creature with "gray, scaly skin" was reported by fisherman Charles Bungay in Fortune Bay in May 1997. The monster had a horselike head on a nearly six-and-a-half-foot-long neck.

Another fisherman said he saw what looked like a dinosaur in Bay L'Argent in the early 1990s.

Kraken

Legends of squid so large it can entangle sailing ships and drag them beneath the waves have existed since man sailed the seas. Those legends are taken a bit more seriously in parts of Newfoundland and Labrador because tentacled monsters of nearly Kraken size have been seen on its shores.

A squid measuring twenty feet washed onto the shores of Portugal Cove, Newfoundland, in 1873 and was photographed by Rev. Moses Harvey of St. Johns, but that specimen was dwarfed by one that washed up on the shores of Glovers Harbour in 1878—it was nearly fifty-six feet long, according to the November 26, 1949, edition of the *Illustrated London News*. Fifty-six feet is nearly as long as a bowling lane.

The existence of such creatures in the area appear in the work of Danish zoologist Japetus Steenstrup who recorded evidence of squid of enormous size from whalers who retrieved jaws, tentacles, and eyes from the belly of sperm whales. Science has discovered a large species of squid—the giant squid—to live off the coast of Newfoundland. These squid grow to approximately forty-six feet long.

Adlet

The Inuits of Newfoundland and Labrador spoke about the Adlet, the offspring of a woman who had sex with a gigantic dog. The Adlet appears human but has dog legs, which allow them to run as fast as dogs.

ADLET

The woman gave birth to ten puppies, five of them (as legend has it) ran across a great sheet of ice all the way to Europe and became the first Europeans. The other five fell into depravity and fed upon the Inuits.

Nova Scotia

Nova Scotia, one of the four provinces of the Atlantic Canada region, is small. Really small. With an area of 21,345 square miles (which includes nearly 4,000 islands), it is the second smallest province, just after Prince Edward Island. To put it in perspective, Nova Scotia is about the size of the state of New York with a population (nearly 950,000) close to that of San Jose, California. The province is within the Appalachian Mountains and features low mountain ranges and hills, along with forests, lakes, barrens, and seacoast. The province is known for its fossils, the rock group April Wine, Samuel Edison (the father of Thomas Edison), singer Anne Murray, and lots and lots of sea creatures.

Sea Monsters

Sea monsters are not just tales from the imaginations of superstitious sailors. Encounters with these aquatic beasts have been reported by motorists, vacationers, and normal working people, like this 2003 encounter a lobster fisherman reported near the lighthouse at Point Aconi, Nova Scotia. Wallace Cartwright saw what he thought was a floating log until one end of the log rose from the water—and it had a head.

"It looked to me like it was a brown animal. It had a head... something shaped like a sea turtle," Cartwright told CBC Radio's *As It Happened* program. "And it had a body on it like a snake, and the girth on the body would be something about the size of a five-gallon bucket."

The brownish snakelike creature, about twenty-six feet long, eventually dipped back into the water and disappeared.

"I have been lobster fishing for thirty years. This was one distinct animal," Cartwright told the CBC. "One I've never seen before."

Stories of sea monsters in the waters of Nova Scotia date back to First Nations peoples. Zoologist Andrew Hebda, curator of zoology at the Nova Scotia Museum of Natural History, wrote in his book *The Serpent Chronologies: Sea Serpents and Other Marine Creatures from Nova Scotia's History* that the local Mi'kmaq people carved images of these beasts into stone.

"If you go down to Keji you can actually see three petroglyphs that clearly have sea serpent motifs," Hebda told the CBC.

The first sea monster encounter by a European in Nova Scotia was recorded by Irish monk St. Brendan who travelled to Canada in the sixth century. He wrote in his diaries of a water beast so enormous he and the other explorers didn't recognize it as a living creature—at first. The group landed on an island that turned out to be Janconious, a legendary monster of the region.

Hebda told the CBC there have been 31 sea monster reports from Nova Scotia during the past 140 years. He dismisses most of them as misidentified basking sharks.

The Sea Monster on the Northumberland Strait

Although sea monsters abound in the waters off Nova Scotia, a monster was busy on the Northumberland Strait in the mid-1800s. The Northumberland Strait separates Nova Scotia and New

SEA MONSTER ON THE NORTHUMBERLAND STRAIT

Brunswick from Prince Edward Island. It's between 56 and 213 feet deep and has some of the warmest ocean waters in Canada.

Millwright William Barry fished off a pier in the small village of Arisag, Nova Scotia, in October 1844, when he saw something unusual. An unknown living beast, about sixty feet long and three feet around, swam near the pier. "It had natural humps on the back, which seemed too small and close together to be bends of the body. It moved in long undulations, thus causing the head and tail to appear and disappear at intervals," per the *New Glasgow News*.

A similar report came from Merigomish Harbour west of Arisag the next August. Several people on the beach reported seeing an eighty-foot-long serpent in the strait. "It was dark in color and very rough and it raised its head frequently from the water, and its back was either covered with humps or they were caused by the motion of the body," per newspaper reports. People observed this creature frolicking in the water for an hour. "It withered about continually and would bend its body into a circle and unbend it with great rapidity. It eventually succeeded in getting off into deeper water."

Captain Sampson of the ship Louise Montgomery reported seeing the creature in July 1879, ten miles east of Pictou Island.

Lake Ainslie Monster

Cape Breton Island's Lake Ainslie is Nova Scotia's largest natural freshwater lake. At twelve miles long and three miles wide, this fifty-nine-foot-deep lake has plenty of room for the Tcipitckaam (or *Jipijka'm*—"the horned serpent"), a monster out of Mi'kmaq legends.

Wilson D. Wallis wrote of a man's encounter with a group of the Tcipitckaam of Lake Ainslie in his book *The Micmac Indians of*

Eastern Canada. This man was at the lake with his family when the Tcipitckaam rose from the water.

"The male was black and probably was larger; the other, brown, was the female. The head is shaped somewhat like that of a horse, but is larger."

This water creature often appears as a tiny worm, but when hungry or threatened it can transform into a ferocious monster the size of a bison.

Green Hill Creature

What would a Canada province be without Bigfoot?

In 2003, teenager Myles MacKenzie heard a noise in the woods that rumbled through his body. It came from a living creature, but no living creature he'd ever experienced, per the *New Glasgow News*.

Myles ran cross-country for his high school team and one day when his father, track coach Stephen MacKenzie, drove Myles and some other boys to Green Hill Provincial Park to practice, his life changed.

As the boys ran through the woods, the day suddenly went silent. The sounds of birds, insects, and even the wind had died. Then something screamed nearby.

"It scared us half to death," Myles told the press.

The boys took off toward the top of the hill and the van Stephen had driven them in, but something else ran too, chasing them.

Myles and his friends suspected the beast that followed them was a Sasquatch, but became convinced when he heard the stories of a monster that reappears in the area every fifty years.

An August 5, 1913, article in the *Thorburn Post* reported a monster that drinks milk from cows and raids chicken coops leaving deep fifteen-inch-long humanlike footprints.

Mrs. Ervin MacKay of rural Thorburn told the *Thorburn Post* in 1913, "There's something evil out there. Something big and shadowy and it scares the bejeebers out of me."

Ontario

Ontario is the second-largest Canadian province (fourth-largest when you count the territories) and the most populated, boasting 38.3 percent of the country's people. It's home to Canada's largest city, Toronto (population 2.7 million), and the nation's capital city of Ottawa. The main topographical traits of the province are forests to the north (66 percent of Ontario is composed of forestland), farmland to the south, and water. Lots and lots of water. Most of Ontario's border with the United States is either composed of rivers or four of the Great Lakes. It's also the final resting place of Jack Fiddler. As the story goes, Fiddler was a killer of Wendigos.

Wendigo—Part 2

The Wendigo, a creature feared for centuries across Canada and the northern United States, is a monster that was once a man. The man, during a time of starvation, fell victim to the will of an evil spirit and resorted to cannibalism. A Wendigo appears as a tall, emaciated, two-legged creature with bones pushing against its skin. The beast has glowing red eyes and the head of a deer.

Wendigo killer Jack Fiddler was born Zhauwuno-geezhigo-gaubow in 1839 to the Anishinaabe people in northwestern Ontario. The Anishinaabe were the last Indigenous people of Canada to live under their own rule. He gained the name Jack Fiddler (and his brother the name Joe Fiddler) from white fur traders who gave First Nations people European names.

Jack, a shaman, was revered for his ability to protect tribes from spells and was known to be able to summon animals to him. He was also said to have defeated fourteen Wendigos, some sent by hostile tribes; others were his own people who had succumbed to cannibalism (although a few Anishinaabe who wanted to eat human flesh had not and simply asked to be killed before they did).

To bring the Anishinaabe under Canadian law and convert them to Christianity, in 1907, the North-West Mounted Police were sent into the region to arrest Jack and Joe Fiddler for murder. Jack Fiddler killed himself in captivity. During Joe's trial, a witness to one of the Wendigo killings, Angus Rae, told the court the Anishinaabe people honestly believed in Wendigos and that the Fiddler brothers were the only ones who could rid the world of their evil. Joe Fiddler was sentenced to death in 1909.

Old Yellow Top

When silver miners founded the town of Cobalt in September 1906, they discovered the area came with an unexpected resident—a seven-foot-tall apelike creature with a mop of yellow hair on its head. People reported seeing the monster for sixty-four years.

Two prospectors, J. A. MacAuley and Lorne Wilson, spotted the creature in July 1923 while testing their claims for silver. They both thought they saw a bear in a blueberry patch near the town, but it was Wilson who picked up a rock and threw it at the beast. The creature stood upright on two legs and the men saw it was no

OLD YELLOW TOP

bear. It resembled an enormous man covered in black fur except for a blond mane that fell from its head across its shoulders.

Old Yellow Top wasn't through with Cobalt. In April 1947, a woman and her son walked along the railroad tracks toward town to grocery shop when the mother saw what she also thought was a bear. It was not. A monster that walked like a man went across the tracks in front of them taking no notice of the two. She said it was covered in dark brown hair except for a yellow patch on its head.

The last sighting was in August 1970. A bipedal creature walked across the road in front of a vehicle carrying miners. The driver, Amos Latrielle, lost control of the vehicle and nearly drove it off a rise. Like the other accounts, the monster had dark fur except for a head of long blond hair.

Dogman—Part 2

Stories of dogmen—seven-to-ten-foot-tall canines that can walk on two legs—have been reported across North America for centuries. They're still reported and quite a few come from Ontario.

Per the blog of Linda Godfrey, author of *The Beast of Bray Road* and other books about dogmen, a man fishing near Bancroft, Ontario, saw something he couldn't rationally explain on July 7, 2015.

He drove to his cabin in the woods when the headlights of his truck hit what he thought was a man standing on the gravel road. The truck was equipped with extra lights on top and when he switched them on, he saw something that was not a man.

"It was a creature about seven feet tall, black with grayish-silver parts, hunched over with a dead rabbit in its hands," he explained to Godfrey. "Its feet seemed to be bent backward. It turned its head and shoulders and looked right at me. I could see its yellow eyes shine."

The beast growled at the fisherman when he decided to floor it and drive straight at the creature, which simply disappeared—whether by jumping out of the way or ducking flat under the high clearance of the truck, the man didn't know. The creature wasn't through with the fisherman, however. The man said something large prowled around his cabin later that night.

From the website dogmanencounters.com, a woman named Eleanor described her dogman encounter from fall 2014. She drove west toward Hamilton, Ontario, when she saw a dog shoot from the bushes on the side of the road and run out in front of her. This "dog" was at least eight feet long, not including the tail.

The Ugly One

In early May 2010, two nurses walking their dog on the shores of Big Trout Lake near the town of Kitchenuhmaykoosib, Ontario, found the corpse of an unknown animal amateur cryptozoologists on the internet dubbed the new Montauk Monster (it resembled the corpse of a mysterious animal discovered on a beach in Montauk, New York, in 2008). However, legends of the local First Nations people, the Kitchenuhmaykoosib Inninuwug, have the lake home to everything from mermaids to a giant serpent to something that resembles what the nurses discovered. "Our ancestors used to call it the Ugly One. Rarely seen, but when seen, it's a bad omen, something bad will happen," per an article on *ScienceBlog*.

The creature, dragged from the water by the nurse's dog, had a long hairy body with bald patches on its head, feet, and tail, which was "like a rat's tail and…a foot long," per a story in the UK's *Daily Mail*. The nurses, who had no idea what it was, snapped photographs of the beast that made their rounds on the internet.

People speculated the animal could be anything from a raccoon, an opossum, an otter, or, maybe, just maybe, the Ugly One.

Lake Ontario Monster

It sounds like it's right out of a Godzilla movie. The Seneca Indians of Lake Ontario had a legend of an enormous creature, the Gaasyendietha, that lived in the lake. It was shaped like a giant serpent with massive teeth—it could also fly and spit fire. Oh, and the Gaasyendietha's favorite meal was people.

The Seneca weren't the only ones to see this monster. Per the August 14, 1829, issue of *Kingston Gazette and Religious Advocate*, European immigrants claimed to have seen a "hideous water snake, or serpent, of prodigious dimensions" near what is now St. Catharines close to Niagara Falls. The newspaper went on to claim "there can be no doubt of the existence of such monsters in our inland seas."

On July 1, 1833, the captain of the schooner Polythermus said he saw a 174-foot blue serpent in the lake, and in 1842, boys playing near Gull Beach claimed to have seen something similar, although brown and somewhat shorter at thirty-nine feet.

Other reports came from 1872 to the 1930s, but they seem to have tapered off over the decades.

Similar stories of monsters in lakes throughout the region are too numerous to mention—with the exception of Igopogo.

Igopogo

A monster lives in the waters of the 135-foot-deep Lake Simcoe in southern Ontario (about forty miles north of Toronto). The lake, the fourth largest that rests entirely in Ontario, has 160 miles of shoreline and is the perfect size for a creature. Igopogo, named after the more famous lake monster Ogopogo of British Columbia, is shier than its western cousin; only a handful of sightings are on record. Sightings became so rare that until a sighting in 1991, the people who lived around the lake thought Igopogo had died.

It is described as looking like a seal, but one that is sixty feet long. Its face is like a dog with bulging eyes. The body has multiple dorsal fins and the tail of a fish.

To the area's First Nations tribes, Igopogo is known as an invisible, but noisy, god. Igopogo is also known as Kempenfelt Kelly, named after a deep bay in the lake, and Beaverton Bessie by the city of Beaverton.

Some of the more famous sightings by Lake Simcoe-area residents occurred in 1952 and 1963. A study of the lake resulted in a "sonar sounding of a large animal" in 1983, per LiveScience.com. A monster hunter captured what he claimed was video footage of Igopogo in 1991. In 2005, Discovery Canada's program *Daily Planet* attempted to flush out the monster using a boat with an underwater camera and sonar, although the crew didn't find anything out of the ordinary.

Toronto Tunnel Monster

Legends of hairy, child-sized humanoids that live in the waterways of Ontario stretch deep into the history of the Algonquian tribes. When the city of Toronto was built over numerous streams and rivers, they think these water spirits, the Memegwesi, may have become hidden beneath the city as well. Those streams and rivers still exist, running through the sewers and culverts under Toronto.

Although the Memegwesi were known to be easygoing creatures, when offended, they would steal objects from the tribe or send canoes adrift. And what would be more offensive than being buried?

In 1978, a Toronto man reported an encounter with one of these creatures, per the March 25, 1979, *Toronto Sun*. The man, who would only be identified as Ernest, went looking for missing kittens when he crawled about six feet into a culvert. "I saw a living nightmare that I'll never forget," he told the *Sun*. The creature was "long and thin, almost like a monkey, three feet long, large teeth, weighing maybe thirty pounds with slate-gray fur." Its eyes glowed orange. The thing hissed at Ernest, telling him to "go away" before it ran deeper into the darkness.

Ernest's wife, Barbara, said her husband came back to their apartment visibly shaken. "I believe Ernie saw exactly what he says he did."

When *Sun* reporters investigated Ernest's claim, they discovered the culvert emptied into the city's sewer system—and they found a dead cat.

Prince Edward Island

Prince Edward Island is the smallest Canadian province at 2,195 square miles. It could fit into Canada's second-smallest province, Nova Scotia, about nine and a half times. It's not just the size that makes it small, it's the population. Only 146,283 people live on the main island and 231 minor islands that make up the province. That's roughly the population of the city of Barrie, Ontario. Prince Edward Island is one of the three maritime provinces in the Gulf of Saint Lawrence and is separated from the other two (New Brunswick and Nova Scotia) by the 140-mile-long, eight-to-twenty-seven-mile-wide Northumberland Strait. The main island is known for its farmland and produces 25 percent of Canada's potatoes. It also has a monster or two.

West Point Serpent

The Indigenous Mi'kmaq peoples warned the first Europeans who came to Prince Edward Island of a gigantic snake that swam "upon the water" offshore of the big island. It didn't take long for the settlers to encounter the serpent, a twelve-to-seventy-nine-foot-long tubelike monster with short, reddish-brown dark fur and the head of a horse. Its body was made of humps that rose and fell in the water as it moved.

Sightings have continued to modern times. Local woman Carol Livingston told Julie V. Watson, author of the book *Ghost Stories and Legends of Prince Edward Island*, that her father and great uncle were fishing from a boat near the West Point Lighthouse in 1980 when a sixty-to-eighty-foot serpent approached them. The creature simply raised its horselike head from the water, stared at them briefly, and swam away.

Per an article from the *Island Press Limited* newspaper, nine people reported seeing the creature between July and August 1992.

One of the last sightings of the monster came in 2002 when Allison Ellis, his grandson and great-grandson, rode ATVs to the beach at West Cape when something rose out of the still ocean waters.

"There was no ripple in the water," he told *Island Press Limited.* "Just outside the bar there was a head sticking out of the water about two feet or so."

They watched the sea monster swim out to sea, creating a wake like that of a boat.

"I'm eighty-three years old and never seen or heard anything that could explain what I saw then," Ellis said.

SLUAGH

The Sluagh

When people settle in new lands, they bring part of their homeland with them in the form of legends. When Scottish settlers came to Prince Edward Island, they brought with them stories of the Sluagh (nearly 40 percent of the population of the province is of Scottish descent, per the *National Household Survey*).

The Sluagh is a large flock of blackbirds that can carry away unsuspecting people. To make this flock more ominous, in each bird was the spirit of a sinner.

The flock could pluck people from their homes at night if the windows were left open, or more often if a person was walking alone outside in the dark. Some victims of the Sluagh never returned, but others did. These would wake as if from a trance in the woods miles from home with no idea how they got there.

Many such encounters were near the town of Bayfield in a swampy area the demon birds were said to inhabit. Although there are few modern encounters, some longtime area residents remember stories of locals carried away by the Sluagh.

Fairies—Part 2

The Scots not only brought with them stories of the Sluagh, fairies seemed to have come with them from across the Atlantic as well. A tree on Fairy Hill in Gowanbrae, just a spot on the map, was thought to be the home of fairies on the island. However, people visited Fairy Hill with a warning—these fairies weren't Disney, they were wicked.

Fairies weren't limited to Fairy Hill; they lived in the woods on Prince Edward Island, their lilting voices and laughter often heard when no one else was around, luring more than one person to their doom. They were also known to torment families in the area, sometimes beckoning children deep into the forest never to return.

Per a January 7, 1983, article in the *Island Magazine*, the fairies were also capable of stealing a human child from its crib and leaving an ill, weak changeling in its place.

The article references a young family named Kelly who had a beautiful baby boy they left inside one day while Mrs. Kelly went to help her husband in the fields. When they returned, the once beautiful, robust baby seemed pale and sickly. The baby grew, but never regained its health; neighbors began calling it Kelly's Fairy, convinced it was a changeling. The child's behavior, often angry and violent, did little to dispel the rumors. When the boy died at nineteen, the townspeople buried him at night in the cover of darkness.

Prince Edward Island Bigfoot

Although Prince Edward Island is, at its closest, eight miles from the mainland, there are stories of Bigfoot on the island. Musician and Prince Edward Island native Nathan Wiley claimed a film crew he was with encountered a Bigfoot while shooting a short movie on the island in 2005.

As the movie's villains were running across a clearing, something unexpected ran with them—a man-sized furry figure on four legs that stopped and stood on two legs before it walked into the woods. However, things happened too fast for most of the crew to get a good look at the creature.

"The only one there that got any kind of a real look at it was the camera guy," Wiley told the website the *Endangered Left*. "The rest of us had to wait to check out the footage."

Since the largest wildlife on Prince Edward Island is the eastern coyote, whatever the crew saw wasn't a native animal. It's also unlikely enough a Bigfoot is roaming the fields and forests of Prince Edward Island; the blurry video has been labelled a hoax.

"I never rule anything out as a hoax," Wiley told the *Endangered Left*. "There are a lot of things that we don't understand; I just don't get too hung up on naming them."

Quebec

For bordering three provinces (Newfoundland and Labrador, New Brunswick, and Ontario) and four states (Vermont, New Hampshire, New York, and Maine) Quebec is just as surrounded by water. The French-speaking province touches Hudson Bay, James Bay, Ungava Bay, Hudson Strait, and the Gulf of Saint Lawrence. The Saint Lawrence River (which connects the Atlantic Ocean to Lake Ontario) cuts through the southern part of the province, touching three of the province's largest cities (Montreal, Quebec City, and Trois-Rivières). Quebec is the largest province in total area at 595,391 square miles (although smaller than the Nunavut Territory). Although Quebec has the second-largest population of all provinces at 8,164,361, most live along the Saint Lawrence River, leaving the rest of the province lightly populated by people, but filled with trees, lakes, rivers, mountains, and monsters.

Memphre

Lake Memphremagog stretches 32 miles south from the city of Magog across the US border and slightly past the Vermont town of Newport. The lake is long—longer than the lake home to Scotland's Nessie (Loch Ness at 22.5 miles), but not nearly as deep. The maximum depth of Loch Ness is 745 feet; Lake Memphremagog's maximum depth is 351 feet, about the length of an American football field. This may be plenty deep enough for a monster named Memphre.

When European settlers arrived in the area, the First Nations people warned them about a huge serpent that lived in the depths of Lake Memphremagog. The first official sighting by settlers was in 1816.

Memphre researcher and self-proclaimed "dracontologist" Jacques Boisvert, who died in February 2006, made more than seven thousand dives into the lake looking for the monster, but he never found it. His research into Memphre uncovered many references to the monster in the local press, including a January 21, 1847, article in the *Stanstead Journal* in which an eyewitness said, "I am not aware whether it is generally known that a strange animal something of a sea serpent…exists in Lake Memphremagog."

Sightings continue to this day and usually describe the creature as having a hump and a long, slender neck topped with a horselike head.

The locals have taken advantage of having a famous monster in its midst; narrated monster tours are offered aboard the *L'Entregens II*, a twelve-passenger pontoon boat. Patrick Corcoran, who worked for Tours Mempremagog in 2011, told Canadian Television he's not convinced there's a monster in the lake.

"The fact that the lake is 360 feet deep in two sections, there's a good chance there is a large fish, because that's where it would be," he said.

In 2011, the Royal Canadian Mint honored Memphre by putting the creature on a full-color collector quarter in its Mythical Creatures series. And, yes, the mint gave Bigfoot some love as well.

Champ

Lake Champlain, a 304-square-mile, 400-foot-deep freshwater lake, is mostly in New York State, but goes into Vermont and Quebec.

It's also the home of Champ, one of North America's most famous lake monsters.

Like the First Nations people who lived near Lake Memphremagog, the local inhabitants of the area surrounding Lake Champlain warned early white settlers of the monster. The first newspaper account was from July 1819, when Captain Crum of the ship *Bulwagga Bay* saw a 187-foot-long black monster swimming nearby, per the *Plattsburgh Republican*. The monster had a head like a horse, which stuck out about fifteen feet from the water. The monster was also seen by a railroad crew, a county sheriff, vacationers, and fishermen.

Ponik

Monsters aren't just purported to swim in big bodies of water. The five-and-a-half-mile-long, 135-foot-deep Pohenegamook Lake on the border of Maine has its own stories.

The first report of the monster of Pohenegamook Lake was in 1874, but sightings were rare until dynamite was used to renovate Route 289 during 1957 and 1958. The explosions apparently dredged something from the depths. Many people have reported seeing the creature, which is generally described as a dinosaur-like beast with four legs, a long neck and tail, although some have said it looks like a crocodile and others a manatee. Those who've claimed to see Ponik include lumberjacks, children, and a priest.

Researchers travelled to the town of Pohenegamook in 1982 to discover more about the creature. "We weren't interested in finding lake monsters. We were interested in the sightings of lake monsters by people," Claude Gagnon, a University of Quebec philosophy professor told *UPI*. "But when you compile all the evidence, you realize there must be something there because the stories are all so similar."

The professor and French writer Michel Meurger spent six months in the province doing research for their book *The Monsters of Quebec's Lakes.*

Documented sightings of Ponik in modern times include a dark figure whose large dorsal fin broke the water in 1974, and researchers using sonar discovered a living creature about twenty-six feet long moving under their boat.

Inhabitants of the town of Pohenegamook named the monster Ponik in 1974.

Loup-Garou

Hunters and trappers beware. Stalking the deep forests of the province may be a shape-shifting monster that sometimes walks the earth as a man and other times in the form of a wolf.

Per a 1973 story on the CBC radio program *Quebec Now*, in the early days of Canada, a man named Léo camped with men he did not know, Hubert Sauvageau and André. While Sauvageau went off into the trees, a werewolf—a Loup-Garou—thundered into the camp and attacked Léo. Reacting swiftly, André threw his good luck charm at the monster, which cut the beast's head. The Loup-Garou's body thrashed and changed before them, turning from a humanlike wolf into Sauvageau. André's attack apparently freed Sauvageau from the Loup-Garou curse.

Forests weren't the only places to be mindful of attacks; in the mid-1700s, people wandering Quebec's city streets were also warned to be on their guard.

Accounts of a Loup-Garou terrorizing the province began in July 1766 and lasted until December 1767. The July 21, 1766, *Quebec Gazette* reported a beggar seen in the Saint-Roch neighborhood who would attempt to persuade passersby to follow him, only to transform into an animal and attack. "This Beast is said to be as dangerous as that which appeared last Year in the Country of Gevaudan; wherefore it is recommended to the Public to be as cautious of him as it would be of a ravenous Wolf," as per the newspaper account.

The *Quebec Gazette* reported on December 11 of that year the beggar-turned-wolf man also caused mischief in the nearby town of Kamouraska. "Beware then of the Wiles of this malicious Beast, and take good Care of falling into its Claws," per the newspaper account.

Wemindji Monster

Sightings of Bigfoot are synonymous with Canada. Although encounters in Quebec aren't as common as they are in the western provinces, the sparse population in its northern reaches probably accounts for that. However, there are sightings, such as this 2013 encounter by a hunter of the Cree community of Wemindji near the shores of Hudson Bay.

Melvin Georgekish saw a pair of red eyes glowing at him from the trees as he drove through an area of woods near Wemindji, according to the CBC. He thought about those eyes as he drove and realized no animal in that area had red eyes, or stood that tall. He turned the truck around and went back to the spot, but the creature had gone. "I am a hunter and I've never seen something like that," he told the news agency.

The next day he returned to the spot and found enormous footprints pressed into the thick mossy ground cover. One was twenty centimeters long, the other thirty-five centimeters. "You can see the toes, too," he told the CBC. "It's like a human foot, but way bigger than a human foot. Wider, too."

CHENOO

Chenoo

The giants of Quebec take hangry to another level.

The Indigenous Algonquin peoples have a legend of a tribe named the Chenoo who angered a Devil. This Devil cursed the Chenoo, turning them into a race of fanged giants with the taste for human flesh. The Chenoo become so hungry, they chew off their own lips; the hungrier they are, the bigger they get.

Of course, with legends, there are multiple stories.

Some stories have Chenoos as men who became giants after eating the meat of a human (like a Wendigo) or is transformed when an evil spirit comes upon him. Either way, the man's heart turns to ice as he becomes a monster so wicked even the scream of a Chenoo will kill anyone who hears it.

How to avoid a Chenoo? Stay out of the woods.

How to kill a Chenoo? This is a little more fun: 1) Get it to vomit out its heart. 2) Get it to swallow enough salt to melt its heart. 3) Chop them into little bitty pieces, and, oh, I don't know, feed them to Memphre, or something.

Gollum

Something Tolkienesque may be stalking the forests of Quebec. In August of 2024, Audrée Tanguay Fréchette was shooting video of a moose when a white, nearly six-foot-tall spindly humanoid crept into view. Fréchette said it was apparently stalking the animal.

"I was filming a moose on a roadside in Gaspésie, Québec, Canada," Fréchette wrote alongside a post of the video on YouTube. "Looking at the video I saw this strange shape at the back left. Can someone tell me what it is?"

It resembles the wicked creature Gollum from the J. R. R. Tolkien's novels *The Hobbit* and *The Lord of the Rings*.

Look at the video "Woman Sees Mysterious 'Gollum-Like' Creature in Canadian Woods" on YouTube, recorded on July 29, 2018, and decide what this thing might be.

Saskatchewan

Saskatchewan is a long strip of a province wedged between Alberta and Manitoba like the stripe of vanilla in a block of Neapolitan ice cream. It's not as bland as vanilla, though. With cities named Saskatoon, Moose Jaw, and Oxbow, how could it be? Ten percent of its 251,700 square miles is filled with rivers and lakes (100,000. Seriously, there are 100,000 lakes in Saskatchewan). The rest of the province is covered by prairie and forests. Called the Land of Living Skies, Saskatchewan is known for its grain farms and is the world's largest producer of potash (salt rich in potassium). One thing the province doesn't have is population; just over a million people live there. There's also something lurking in Turtle Lake.

Turtle Lake Monster

Turtle Lake is a long, thin body of water in west-central Saskatchewan; about thirteen miles long and three miles wide. Although relatively shallow, forty-seven feet at its deepest, legend has it is home to an enormous fishy critter.

The local First Nations Cree tribes that lived around the lake warned settlers of Big Fish, a creature ten feet to thirty-three feet long with no dorsal fin and the head of a dog. Big Fish was known to threaten boaters and destroy fishing nets. People venturing far out into the lake would often disappear.

Almost every year, fishermen claim to see something large in the lake swimming next to their boat. According to *Prince Albert NOW*, local resident Charlie Kivimaa is one of them.

"I was at the boat launch, and I saw this thing coming toward the shore. It looked like a great big fish, like a whale; then it went down," Kivimaa told *Prince Albert NOW*. "I would say it was probably ten or fifteen feet long."

A spokesperson from the Ministry of Environment said the monster may simply be a river sturgeon, even though river sturgeon can only grow to be about ten feet long.

Reindeer Lake Monster

This 2,568-square-mile body of water on the border of Manitoba is the second-largest lake in Saskatchewan and ninth largest in Canada. It would be a shame if it didn't have a monster story.

It does.

The lake is supposedly named after the caribou that would come this far south during the winter migration. The First Nations Cree tribes that inhabited the area for a millennium have stories of a monster living in Deep Bay, a circular, 722-foot-deep impact crater formed during the Cretaceous period. The monster that dwells in that crater would eat reindeer that fell through the ice.

In 2006, the CBC reported a mystery creature in the lake had gutted three sled dogs on an island; other dogs were missing. Northlands First Nations Chief Joe Dantouze told the CBC people blamed the deaths on an unknown animal.

"There were signs that people had seen something out on the lake but they don't know what type of animal it would be," Dantouze said in 2006. "They're afraid right now. People are afraid to go out on the lake, like for fishing, and to use the beaches around the lake."

What could the monster be?

Kootenay author Art Joyce told the *Nelson Star*, "That whole area of northern Saskatchewan is dotted with lakes, all over the place, that are the result of the retreating glaciers. Who knows? Perhaps some relic dinosaur could exist in the deep water, deposited there at the end of the last ice age."

Zoobey

Saskatchewan wouldn't be able to claim any bragging rights among the provinces if it didn't have at least one Bigfoot sighting.

The town of Rockglen (population four hundred) in the south-central part of the province has its own big hairy beast known to the locals as Zoobey, Zoobie, or sometimes just Zoobs. According to the *Assiniboia Times*, this ten-foot-tall apelike creature smells like a skunk and can emit a piercing scream. Often seen near Columbus Hill in the town, this monster has been known to scream and chase the local wildlife. There have been at least four sightings of Zoobey since 2009.

Dire Wolf

In 2015, a fishing guide in northern Saskatchewan encountered an enormous wolf, which he recorded. His dog was lucky to survive.

In the video, a black figure lies waiting in the grass in a patch of thin young trees as a barking dog approaches it. The dark animal leaps from the grass and rushes the dog; it is an enormous black wolf the website Unilad claims is nearly eight feet long. The wolf chases the dog across the screen and the dog yelps before the great beast slinks back into the woods.

DIRE WOLF

The Canadian timber wolf is the largest wolf in the world. An adult animal is typically between three and four feet in length, including the tail. Some who viewed the fishing guide's video on YouTube speculate the creature may be a remnant dire wolf. Dire wolves, a Pleistocene animal that survived until 9,440 years ago, grew up to over six feet in length.

According to the *National Post*, there are enormous wolves in northern Saskatchewan and, unlike the normal behavior of wolves, these have been known to stalk and attack humans.

Northwest Territories

The Northwest Territories, north of the provinces of Alberta and Saskatchewan and venturing into the Arctic Circle, is the second-largest territory in Canada (after Nunavut) at 452,480 square miles, and sixth in size when including the provinces. Its population is tiny at 44,291, which is about the size of Montana's fourth-largest city, Bozeman, and the state of Montana's population is nothing to brag about. The territory is composed of coniferous forest and tundra, broken up by mountains and bodies of water, such as the Mackenzie River, the thirteenth-longest river in the world at 2,635 miles; Great Bear Lake, the largest lake solely in Canada with a surface area of 7,474 square miles; and Great Slave Lake, North America's deepest body of water with depths of 2,014 feet. Wood Buffalo National Park, Canada's largest national park, is in the Northwest Territories; it is larger than Switzerland. It's also known for a big doggie.

Waheela

The beautiful Nahanni Valley, nestled in the 6,834-square-mile Nahanni National Park Reserve in the southwestern corner of the Northwest Territories, can only be reached by foot, airplane, or the Nahanni River that rushes through it. There are hot springs, waterfalls, and white-water, which makes the valley popular with rafters, hikers, campers, and sportsmen.

The Nahanni Valley, however, is also known by a grimmer name—the Valley of the Headless Men.

Two brothers out to discover gold, Willie and Frank McLeod, walked into the Nahanni Valley in 1908 and vanished. They were finally found in 1917 on the riverbank, headless, per an article in the *Outdoor Journal*. That year Swiss prospector Martin Jorgenson embarked to the Nahanni Valley, sending letters home that he'd found gold. When word from Jorgenson stopped, a party set off to find him; his headless remains were among the ashes of his cabin. Trapper John O'Brien was found in the valley not long after, missing his head. In 1945, an Ontario miner was found headless. He was still in his sleeping bag.

How they lost their heads wasn't a mystery. The local First Nations tribes knew—it was the Waheela.

Waheelas are monstrous bearlike wolves—more than three feet tall at the shoulder—like the dire wolf that went extinct 9,440 years ago. However, the head of the Waheela is larger, its toes wider, and its front legs longer than its rear legs. The beast has long white fur and could just be really, really friendly. You know, overly aggressive licking, or maybe biting people's heads off. The usual.

Ch'ii Choo

There's a spotting of lakes in the Northwest Territories that look like footprints. Giant footprints. Aerial pictures of the lakes/prints sent to CBC North do, in fact, look like enormous human footprints. It's always fun to find natural objects that are supposed to look like something else, like the Dead Man of Ireland, the Old Man of the Mountain in New Hampshire, and the Face on Mars. (Unless, of course, it's really a face. Hmm?)

These lakes, however, have a connection to the First Nations legend of a traveler who went by different names in different tribes: Atachuukaii, Yamoria, Zhamba Deja, Hachoghe, and Yamozha. But it doesn't matter if the names are different—the story is the same. This traveler was also a warrior who defeated a man-eating giant named Ch'ii Choo.

This traveler battled the giant, chasing him from what is now Fort Good Hope to Norman Wells. The giant Ch'ii Choo's footsteps left indentations in the ground large enough to become the lakes.

Giant Beavers

Ch'ii Choo wasn't the only giant the traveller battled. Giant beavers often attacked tribes that lived near Great Bear Lake. The hero of the Dene called Yamoria chased the beavers down the Mackenzie River and killed them. After skinning them he cooked the beavers; the animals' fat dripped into the fire and started smoking. The smoke can sometimes be seen to this day.

Another story has Yamoria running the beavers into what is now Saskatchewan where the beavers destroyed all the trees in the area during their battle and left the Athabasca sand dunes. Yamoria created an island during the battle when he threw the beaver's dam into the Athabasca River.

GIANT BEAVERS

Giant Wolverines

Enormous wolverines were apparently also a problem in the early days of the Northwest Territories. A giant wolverine destroyed the land around the Tsiigehnjik River tearing apart rocks and dirt to dig a den. Monstrous wolverines also attacked the Dene. Yamoria killed them and squeezed their offspring until the young wolverines were the size of the animal we know today.

Nàhgą—The Tlicho Sasquatch

Large, hairy, magical humanlike creatures have stalked the Tlicho region for millennia, stealing silently into backcountry camps to kidnap the unsuspecting.

Per the CBC, Tlicho Elder Michel Louis Rabesca said his people and the Nàhgą have always lived side by side. The Nàhgą sometimes look like people and even wear clothes and can charm those who encounter them and coax these poor souls into the wilderness. No one sees these people again.

Fisherman Tony Williah told the CBC in 2016 he fell into the Lac La Martre, the third-largest lake in the territory, and had a horrifying encounter with a Nàhgą. He'd attempted to pull trash from the lake and tipped over. Unable to climb back into his boat, he made his way to the nearest shore.

"I managed to swim to an island at the end of the point," he told the CBC. "All of a sudden, there was a big man standing beside me."

When he tried to get a better look at the "man," the figure was gone, although heavy branches snapped throughout the bushes behind him.

At this point, Williah, terrified, went back into the frigid water and swam to the mainland. The next two days he wandered in the forest, wondering if the thing he'd seen on the island had followed him. The Royal Canadian Mounted Police and the Canadian military found Williah and took him to the hospital.

He is convinced he encountered a Nàhgą.

Nunavut

If you want to travel to the ends of the earth without visiting Oymyakon, Siberia (the coldest inhabited place on Earth with an average temperature of 4.1 degrees Fahrenheit), the Canadian territory of Nunavut would be a close second. Nunavut, the largest Canadian territory at 808,185 square miles (the size of Saudi Arabia), has a population of roughly 38,000 people, most of whom are Inuits. The territory's largest (and only) city, Iqaluit, has a population of 7,740. Is it cold in Nunavut? Sure. The average high temperature for the village of Eureka in the northern part of the territory is just .5 degrees warmer than Siberia. Nunavut includes islands, rivers, mountains, forests, arctic tundra, and the Qalupalik.

Qalupalik

A green-skinned humanoid creature with long hair and knifelike fingernails lies in wait under the waters that surround Nunavut. The Qalupalik lives in the sea, watching and waiting for a child to wander close to shore, then it attracts the child to the water with a hypnotic hum. It grabs the child, tucks it into an amautik (an Inuit parka with a baby pouch), then spirits it off to sea to keep it forever, per the *Anchorage (Alaska) Daily News*.

Frank Boas, in his book *The Central Eskimo*, relates the story this Inuit legend is based upon.

QALUPALIK

A woman who lived with her grandson was so poor the two rarely had food. The boy cried so much from hunger his grandmother summoned a Qalupalik to be rid of him. The creature came, wrapped the boy into its parka, and vanished into the ocean waters. Remorseful for her actions, the grandmother asked hunters in her village to search for her grandson. When the weather warmed and the ice cracked, the boy ventured to the surface to play. When hunters approached him, the boy, who did not want to return to his grandmother, tugged a seaweed rope the Qalupalik had tied around him, and the monster pulled him back into the water. This repeated until the hunters snuck up on the boy, cut the rope, and took him back to his grandmother.

This story kind of makes me feel sorry for the Qalupalik.

Angeoa

An ancient beast lurks beneath the surface of Dubawnt Lake near the border of the Northwest Territories. At 774 feet deep, the lake, with a surface area of 1,402 square miles, is large enough for the Angeoa, a creature about the size of a sperm whale. The local Inuits claim the Angeoa is a hostile black beast with an enormous fin; it flips boats and eats the fishermen.

Canadian environmentalist Farley Mowat spoke with an Inuit man in the 1940s who claimed the Angeoa overturned his father's kayak in the late 1800s. The monster ate the second man in the boat.

Mahaha

Giggles should be reserved for clowns, but then again, clowns are scary too. The Mahaha of the north is a thin blue-skinned being that roams mostly naked through the snow, glaring through pure white eyes for human victims. Oh, and giggles flow from its permanent Joker smile—it always giggles.

The legend of the Mahaha stretches back centuries. This entity wanders through the deadly cold, barefoot and dressed in rags, its long, greasy hair hanging low on its back. The creature approaches people it has mesmerized with its maniacal giggle and touches them with long, thin fingers. A cold radiates through the victim's body, and the Mahaha begins to tickle them—until the victim is dead. When found, the victims often wear a smile frozen onto their face.

This frozen demon, however, is stupid. Many Inuit tales describe how a person tricked a Mahaha into taking a drink of water, then pushed the creature into the waves.

Ijiraq

This red-eyed shape-shifter is particularly loathsome as it only preys on children. The entity sneaks into villages in the guise of whatever it chooses and lures children into the wilderness where it snatches them and hides them. The children almost never see their families again. The only way these children can escape is if the child is clever enough to talk the Ijiraq into releasing them.

The Ijiraq can be seen straight-on only if they choose. Otherwise, people can see the Ijiraq from the corner of their eye, but when they turn toward the creature, it becomes nothing but shadow. These creatures of shadow are believed to inhabit a place between worlds and are part of each, but also neither. Legend has it the Ijiraq were once Inuit who hunted game too far to the north and became trapped between the worlds of the living and dead.

One other trait of the Ijiraq is confusion. When hunters encounter an Ijiraq, they often see mirages and become forgetful, returning from hunting trips empty-handed although they don't know why.

Bigfoot—Part, Whatever

People have encountered Bigfoot in Canada? Who knew?

Over the years, residents of Sanikiluaq (a small village on Hudson Bay's Flaherty Island) have witnessed a six-to-eight-foot-tall humanlike creature lurking in the shadows. In 2001, the monster grabbed the attention of the Nunavut government.

Per the *Nunatsiaq News*, a government official found enormous humanlike footprints whose gait measured more than three feet. The average human male gait is less than one and a half feet.

"It's definitely not a bear," Weenusk Chief Abraham Hunter said in an interview with Canada's *Nunatsiaq News*. "I looked at them. They were six feet apart, walking."

Yukon

The population of the Yukon territory is 35,874. With an area of 186,272 square miles, that's 0.19 people per mile. Given that the Yukon has one major city—*one*—Whitehorse, with a population of 25,085, that means there's a lot that goes on in the Yukon no one ever sees. Canada's westernmost territory is composed of spruce forests, arctic tundra, rivers, and mountains. Mount Logan is one of the highest mountains in North America at 19,518 feet, second only to Denali (formerly Mount McKinley) in Alaska at 20,308 feet. The territory is bordered by the Northwest Territories, British Columbia, and Alaska. The Yukon is famous for the

Klondike gold rush of the 1890s and the Yukon Quest dogsled race that traverses 994 miles. It's also home to the Saytoechin, or Yukon Beaver Eater.

Saytoechin

A large hairy beast lurks along the waterways of the Yukon, and it's not Bigfoot. Well, there is a large, hairy beast that lurks along the waterways of the Yukon, and it *is* Bigfoot, but this monster is the Saytoechin, a creature the First Nations people claim is larger than a grizzly bear that feeds by destroying lodges and devouring the beavers within.

According to the British Columbia Scientific Cryptozoology Club, a First Nations family fishing at Tatchun Lake in central Yukon in the mid-1980s witnessed an animal "eight or nine feet high, bigger than a grizzly bear…coming toward them." They fired shots to frighten the animal and delayed it enough to get their boat motor started and escape. The animal, with a flat face and long tail, they claim, was a Saytoechin. When shown pictures, the witnesses identified the beast to be a giant ground sloth.

The giant ground sloth *Megaloynx jeffersonii* was native to North America, including the Yukon, from the Late Miocene through the Pleistocene, supposedly becoming extinct eleven thousand years ago. Although sloths are herbivores, some scientists speculate the giant ground sloth may have been an omnivore.

Could a thought-to-be-extinct species of megafauna still be alive in the wilds of the Yukon territory? The two-meter-long, 90-kilogram coelacanth was rediscovered by science in 1938 after thought extinct for the past 65 million years. We never know what's out there.

Partridge Creek Beast

In 1908, French writer Georges Dupuy published a monstrous tale in the French magazine *Je sais tout* (I Know Everything) about an encounter between hunters and a theropod dinosaur near the Partridge River, a water flow that leads from Partridge Lake to Bennett Lake and eventually into the Bering Sea.

Dupuy's story claims James Lewis Buttler and Tom Leemore hunted moose when the animals became spooked and ran. The hunters followed the herd and discovered enormous three-toed tracks among the hoofprints. The three-toed footprints were nearly two meters long and a meter wide. When the hunters returned to civilization, Dupuy, Father Pierre Lavagneux, a French missionary, and five First Nations hunters went back to the area to search for the animal that made the tracks.

They found it. A thirty-foot-long bipedal dinosaur Dupuy said was a Ceratosaurus. This creature, with an enormous skull holding elongated teeth and a horn on its nose, lived during the late Jurassic period.

Years later, the French missionary Lavagneux wrote a letter to Dupuy saying he again saw the Partridge Creek Beast in December 1907 eating a caribou.

Unfortunately, the reality of a remnant dinosaur living in the frigid wilds of the Yukon is remote at best.

The Yukon Camel

The existence of camels in the Yukon is solid and flimsy at the same time. Camels existed, and may still exist, in the Yukon, possibly along with a cameloid cryptid locals call the Ur Chow.

Camels that once lived in the Yukon were the Yukon giant camel and the western camel. The 3.5-million-year-old giant camel was

YUKON CAMEL

the forerunner of today's camels, escaping the last ice age across the Bering Land Bridge and venturing over Asia and into Northern Africa. The western camel, although it lived until more recent times in the Yukon than the giant camel, had a lonely bachelor life and died at home in its apartment. More specifically, it lived during a time the land bridge didn't exist, so it never escaped North America and became extinct.

But that didn't stop the camel from returning to the Yukon by way of the Cariboo Gold Rush. In April 1862, John C. Calbraith purchased twenty-three Bactrian camels to use as pack animals along the Old Cariboo Road, and it turned into a disaster. The other animals on the trail, horses and mules, were terrified of the camels, and the camels themselves were not made for the weather or terrain. Henry Guillod, an Englishman working for Calbraith, wrote in his journal, "Was bothered today by camels of which there are about a dozen here who have a neat idea of walking over your tent and eating your shirts," per the *Ashcroft-Cache Creek Journal.*

The Cariboo Camel experiment was a failure; Calbraith released the animals into the wild, some ending up on plates, the others remaining elusive. The last camel reportedly died in 1905, although stories of camel sightings in the Yukon have continued.

A rare, little-spoken-of cryptid of the Yukon and British Columbia is the Ur Chow. This camel-like creature supposedly roams the forests of northwest Canada, feeding off tree bark it scrapes with the large teeth of its lower jaw.

Wechuge

From the Hudson Bay to the west coast of Alaska comes the Athabaskan story of a creature that is similar to, but not exactly like, a Wendigo—the Wechuge.

Unlike the Wendigo, where a person becomes a gigantic ravenous monster after resorting to cannibalism, a Wechuge is a person who becomes a gigantic ravenous cannibalistic monster after being possessed by an evil spirit.

Depending on the region, sometimes the Wechuge is said to be formed from ice come to life. This makes the creature invulnerable to human weapons—all but fire, which will melt it.

Yukon Howler

In 2004, Marion Sheldon and Gus Jules drove ATVs along the highway near the village Teslin, southeast of Whitehorse, when a legendary howling monster ran across the road in front of them, according to the CBC.

"They claim they thought it was a person standing beside the road, but couldn't tell from all the dust," conservation officer Dave Bakica told the news network. "By the time they turned around to look back, they said this person was completely covered in hair and took just two strides to get across the whole Alaska highway."

They immediately contacted Bakica, but by the time the officer arrived at the spot, rain had erased any possible footprints left by what Sheldon and Jules described as the local Sasquatch—the Yukon Howler. Bakica told the CBC Sheldon and Jules were visibly shaken by the encounter.

Stories of the Yukon Howler stretch back hundreds of years. This six-and-a-half-to-ten-foot-tall, four-hundred-to-twelve-hundred-pound screaming monster that leaves twenty-four-inch-long footprints is as Canadian as the sled dogs it eats. However, given the scant population in the Yukon, only five encounters with the Howler were reported between 1978 and this 2004 sighting.

Chapter 3
The Caribbean

Nearly seven thousand islands dot the 1.063 million square miles of clear blue Atlantic water that make up the Caribbean Sea. From Cuba in the north to Trinidad and Tobago in the south, the Caribbean is composed of thirty-four independent countries, foreign territories, and dependencies lying between the United States, Mexico, Venezuela, and 2,822 miles of open ocean. The islands themselves are composed of beaches, forests (tropical, subtropical, cloud, and coniferous), and, on thirteen of them, mountains. Out of all this, only about one hundred of the islands are inhabited. That allows plenty of room for brutes and beasts.

Andros

Andros is considered the largest island in the Bahamas, although it is an archipelago of three main islands (North Andros, South Andros, and Mangrove Cay) split by estuaries and inlets. Apart from tourism (featuring deep-sea fishing, inland fly-fishing, scuba diving in and around blue holes, and exploring national forests), fruit and vegetable farming makes up the economy of Andros. Home to the third-largest barrier reef on the planet, Andros Barrier Reef at around 140 miles long, the island is popular with snorkelers. The island is also the home of the Chickcharney.

Chickcharney

Upon visiting the Andros plantation of his father, Joseph Chamberlain, UK Prime Minister Neville Chamberlain (in office from May 28, 1937, to May 10, 1940) ordered a large swath of forest felled for farming. Part of this deforestation was to an old-growth pine forest that, unbeknownst to Chamberlain (or maybe he knew and just didn't care), was home to Chickcharnies.

Chickcharney are, in some sense, fairy beings that look like owls. They stand three feet tall on three-toed feet and are covered in fur-like feathers. They possess three-fingered hands and brilliant, glaring red eyes. But the most discussed attribute is their head that can rotate in a complete circle. Some legends describe the Chickcharney as possessing the tail of a South American monkey that allows them to climb trees, according to the *Northend Agents* newspaper.

The Chickcharnies on the Chamberlain land apparently cursed the plantation, and it fell into financial ruin.

Angering one of these creatures can also become deadly. Mischievous by nature, if they are threatened or mocked, they may turn on you and, at best, cause bad luck, or, at worst, twist your head backward. If you're respectful to a Chickcharney, they may bestow good luck.

Andros residents are still cautious of offending the Chickcharney and carry flowers or colorful cloth to appease them.

CHICKCHARNEY

Some believe the legend of the Chickcharney may be based on a flightless burrowing barn owl (Tyto pollens) that lived on Andros during the Pleistocene epoch. Although this species of three-foot-tall owl has been extinct at least eleven thousand years, could a small pocket of the bird remain on this green island, 11 percent covered by forest? Stranger things have happened in nature.

Woo.

The Lusca

This Caribbean monster is right out of a made-for-TV movie.

Seriously. It is.

When 2010's *Sharktopus* premiered on the Syfy network, most viewers probably didn't know New Horizons Pictures based the monster on a Caribbean legend. The tagline of *Sharktopus*, "Half-Shark. Half-Octopus. All Terror." accurately describes the Lusca, a, um, well, a half-shark/half-octopus. (Although there are no legends of the second and third monsters in the sequels, 2014's *Sharktopus vs. Pteracuda* and 2015's *Sharktopus vs. Whalewolf*. Which is a pity.)

Islanders blame this enormous entity for the vanished scuba divers daring to explore the over two hundred oceanic blue holes (there are more than one hundred seventy inland blue holes that are Lusca free). Blue holes are underwater caves formed by glacial run-off during the Pleistocene epoch; as the worldwide glaciers melted, the holes near the shores flooded with seawater. They are named for the deep blue color and drop up to 984 feet. And, although the holes are home to species that live in the surrounding waters, they are the perfect hiding spot for this beast of tentacles and teeth.

Legends tell of Luscas growing more than seventy-five feet long, which is longer than the largest whale shark on record at sixty feet. Other legends put the hostile monster at closer to two hundred feet in length, which is much longer than the thirty feet of the largest octopus, and the forty-three feet of the largest giant squid.

So, if you're a confident diver, go ahead and brave the blue holes with their gorgeous stone walls, colorful sea creatures, and gripping tentacles and gnashing teeth of the terrible Lusca.

If you're not quite up to being eaten, simply visit the Lucky Lusca Tavern on the beach at Staniard Creek and order a conch salad or grilled lobster. It has outdoor seating and, according to Yelp, it serves great cocktails. Oh, and go on karaoke night.

Antigua and Barbuda

The independent Commonwealth of Antigua and Barbuda is composed of two main islands and a smattering of smaller ones at the point where the Atlantic Ocean becomes the Caribbean Sea. Musician Eric Clapton, actor Pierce Brosnan, and talk show host Oprah Winfrey like it so much they own homes there. The nation was colonized by the British Empire, gaining its independence November 1, 1981. It also has this demon...

Jumbee

This spirit is more than just a ghost. The dark, malevolent Jumbee is the demonic spirit of a person who was evil in life and cannot rest in peace, so it inflicts agony upon the living. The foul spirit emerges from the body three days after death and begins its mischief. The Jumbee appears, not as a Western ghost, but more as a shadow person, occasionally shape-shifting into a cat, pig, or dog. These spirits are known to sneak into homes under the cover of darkness and strangle the families inside or break their necks. During ceremonial dances—Jumbee dancers driven by Jumbee drums—the entities have been known to possess one of the eight participants.

JUMBEE

Friendly, they're not.

The Dutchman Jumbee (called so due to the Dutch colonization and continued presence in the Caribbean) inhabits trees. If someone damages a Dutchman Tree, or climbs one, they can expect injury, or death, in a few days.

The Moko Jumbie (regional spelling) were brought to the Caribbean from West Africa. These people are healers who wear masks and colorful clothing and walk on stilts to represent the African god Moko. The Moko Jumbie appears at ceremonies and carnivals to ward off evil and is said to have walked to the Caribbean across the Atlantic Ocean.

Even though the malevolent Jumbee is a terror of the islands, there are ways to protect yourself from this spirit.

Like the legends of vampires, Jumbees are easily distracted, so a pile of salt, rice, or sand at the front door will keep them busy until they've counted each grain—until sunrise. So, heap those piles high.

Depending on the island, the Jumbee either has backward feet or no feet. Leaving shoes outside the house will busy the entity with trying to don shoes it can never wear. Walking backward may confuse a Jumbee.

Like many paranormal entities from around the world, apparently Jumbees can't cross moving water. So, crossing a river or stream should keep you safe. Also, if you're willing to part with a few drops of rum, the Jumbee may be satisfied and leave you alone. A small price to avoid a broken neck.

The Jumbee has roots in the islands of Trinidad and Tobago, the Bahamas, Montserrat, Jamaica, Barbados, and Guyana on the mainland. It is also known in these various locations as the Jumbie, Duppy, Chongo, Mendo, and others.

So, be careful walking home at night, and if you pass a cemetery on the way, as they say in Guyana, "Don't let the Jumbee hold you."

Bay Islands

Lying on the easternmost edge of the Caribbean, the Bay Islands (controlled by Honduras) are a snorkeling paradise. The three islands, Roatán, Utila, and Guanaja, are located along the 621-mile Mesoamerican Barrier Reef that forms an L from Cancún, Mexico, to Belize, to the eastern part of Honduras. The islands boast popular beaches, volcanic crevices for scuba diving, mountains, and whale sharks that visit Utila in April and May. The first Europeans to see the islands sailed with Christopher Columbus. And, does it have monsters? Of course.

Obia

As with many Caribbean monsters, the Obia got its start much farther east, in Western Africa, brought into the Western Hemisphere by the African slave trade. The Obia is an example of why you should never anger a witch.

Created by a witch, this monster can appear as an enormous hairy animal or a small flying creature, depending on the duty the witch wishes it to perform. In Western Africa, the Obia was sometimes conjured to protect the witch's children, although on the Bay Islands, Jamaica, and the mainland, the Obia is usually made to kidnap young girls from local villages and bring them to the witch, who wears the girl's skin (similar to the Gullah culture's witch/vampire Boo Hag legend of South Carolina. See *Chasing American Monsters* for a full account).

The term *Obia* (alternatively spelled *Obeah*), is also used in reference to the witch itself on the Bay Islands, in Honduras, and Jamaica.

LA SIGUANABA

La Siguanaba

Unlike the other creatures that haunt the Caribbean by way of Africa, la Siguanaba is native to Central America. A creature of Maya mythology, la Siguanaba roams the mainland as well as the Bay Islands.

This monster began as a beautiful woman who cared more for her lovers than she did for her cuckhold husband or her children. After trying to kill her husband, the god Tlaloc (god of rain and earthly fertility) punished the woman by transforming her into a beast with a horse's face. She is doomed to forever wander the night in her original beauty to attract men before transforming and terrifying them with her horrendous visage.

La Siguanaba is considered a cautionary tale for anyone who may become unfaithful to their significant other, but she has been used as a figure of female empowerment in Central America.

Cuba

The Republic of Cuba is the largest Caribbean country at 42,426 square miles (about the size of the American state of Tennessee) and is made up of not just the main island, but 4,195 surrounding islands, and Isla de la Juventud (the seventh-largest island in the West Indies located thirty-one miles south of Cuba's main island). The country was ruled by Spain until the Spanish-American War of 1898, then, on New Year's Day in 1959, revolutionary Fidel Castro led a takeover of the country. After declaring itself a communist nation, the United States cut off contact with Cuba, which resulted in, among other things, the vast number of classic American cars in Cuba. Cubans couldn't import American cars for decades, so many 1950s cars are still on the streets. Cuba is also known for producing exceptional rum and cigars, coffee, and art. The island is located at the spot where the Atlantic Ocean, the Caribbean Sea, and the Gulf of Mexico meet. It also has this thing about the Mother of Water.

Madre de Aguas

The horned boa constrictor is enormous; long, and thick as a tree.

The water fairy is beautiful, with eyes as azure as the water in which it lives.

The lovely African woman, whose lower half is a serpent, has long, thick hair, and carries a snake over her shoulders.

Depending on the place in the Caribbean, Madre de Aguas can appear however she pleases. In Cuba, the Mother of Water appears as the horned boa, a monstrous snake with a scale pattern opposite that of the Cuban boa (a huge snake in its own right, growing upward of sixteen feet long). Madre de Aguas lives in the freshwater lakes and rivers of Cuba, protecting the waterways and wetlands of the country and saving them from drought.

This story originated in Africa with Mami Wata (Mother Water) and migrated into the Caribbean with the slave trade. In Africa, Mami Wata means good fortune, but, depending on a person's behavior, can also mean ill fortune.

Do not look into the eyes of the Mother of Waters, or fall ill.

Do not attempt to kill the Madre de Aguas, or die yourself.

Do not…well, don't be a jerk. Be kind to water.

In Cuba (as in Africa), the Mother of Water protects children, and her story is used by parents to keep children away from dangerous waterways.

MADRE DE AGUAS

Grenada

The Carib people (for which the Caribbean is named) never put up with any European BS unless there was liquor involved. Sailors first sighted the Grenada in 1498 during Columbus's third voyage, although they never set foot on the island. The English and French attempted to colonize this island and were pushed away each time by the Carib. However, in 1650, the French offered the Carib knives and brandy in exchange for the island. Then, in predictable European behavior, the new French settlers slaughtered the Carib population. The British gained control of the island in 1783 and granted the country its independence in 1974. Called the Island of Spice because of the cloves, mace, and nutmeg production, Grenada enjoys a mild climate with yearlong highs in the 80s (F) and lows in the 70s. The 133-square-mile island is composed of mountains, jungles, lakes, swamps, plains, and, of course, beaches. It's also home to an ornery creature called Anancy.

Anancy

The trickster god of the Caribbean, the Anancy (also spelled *Anansi*) may appear as a human being or an animal such as a hare or goat, although its true form is that of a giant spider with human features, such as its eyes, nose, and mouth. This god is lazy and uses subterfuge to get what it desires, often to the detriment of those it tricks.

Anancy doesn't only fool humans; it can fool the gods themselves. Although on some islands and in Ghana in Africa, Anancy is sometimes viewed as the creator god, Grenada embraces the spider's trickery. One popular tale is of Anancy fooling the god of the sky to release the rains and the night.

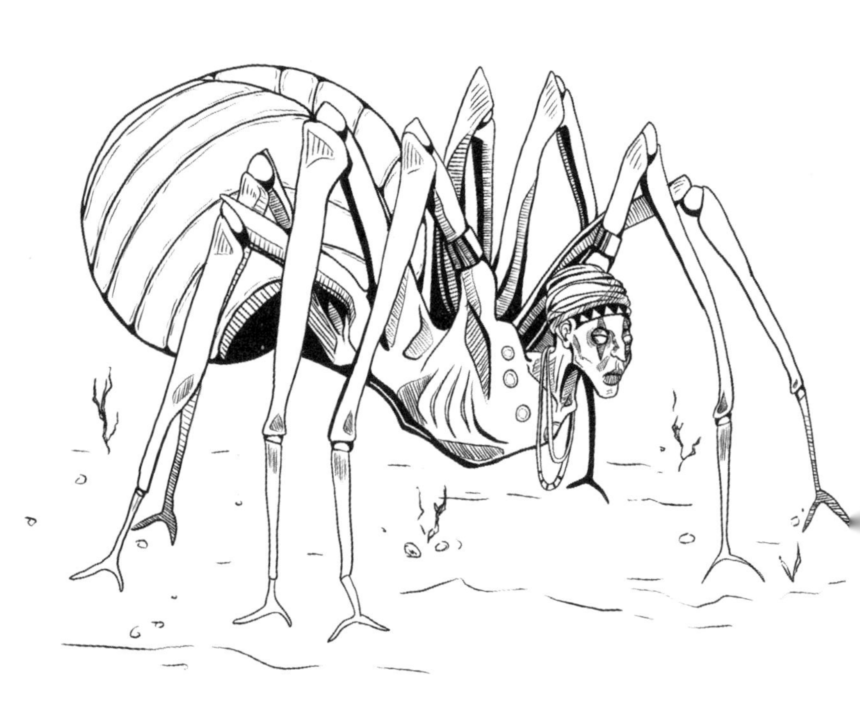

ANANCY

One way to spot Anancy in human form is his speech. The trickster god's voice is high, and he speaks with a lisp. Another way is to beware of the lazy man who avoids work and asks others to do it for him.

Like many other monsters feared, or worshiped, this god came to Grenada and other islands from West African during the era of slavery.

Mamadjo

The Grand Etang is a thirty-six-acre crater lake located in Grenada's Saint Andrew Parish on the eastern side of the island. Grenada's deepest lake, the crystal-clear water of Grand Etang, is home to crayfish, barracuda, bonefish, and freshwater lobster.

It's also home to mermaids.

The Mamadjo, or Mother of Fishes, is considered a water spirit by the locals who, four times a year, gift the Mother with food (including unsalted rice, sheep, fowl, and rum) and play drums and dance for her. To thank the people for their gifts, she provides rain to ensure their crops grow healthy and tall. If Mamadjo were to ever disappear from the lake, the Grand Etang would dry up, and their crops would fail. To keep the mermaid goddess on good terms, no one swims in the lake for fear of angering her.

The origin of Mamadjo is thought to have come to Grenada from West Africa, although mermaids also have their roots in other cultures. However Mamadjo came to the island, one thing is for certain—it wouldn't hurt to bring her rum.

Loogaroo

Like the Loup-Garou of Quebec, the Loogaroo is a shape-shifting monster that sometimes appears as a man, and other times appears as an animal. These monsters from French mythology reflect the French's time on the island. However, unlike the werewolf conjured by French peasants to explain the howls that sent them cowering in their beds at night, there are no wolves on Grenada.

Here, the Loogaroo is a demon that removes its skin and flies through the dark skies as a ball of fire. This entity acts as a vampire, attacking their victims and sucking their blood to either sell to the Devil so they can gain more power or to consume and give the appearance of youth.

Humanoid Fish

A fisherman can never tell what he hooks until hauling it out of the water.

The *UK Daily Mail* reported in 2016, fisherman Hope McLawrence, then seventy-four, caught a twelve-inch fish with wings, a "perfect human nose," and feet with toes in the seas off the island of Carriacou. It had no dorsal fin and an odd tail.

"Quite a few people were pretty scared and thought it looked like something out of a sci-fi horror film," according to an islander interviewed by the newspaper. "I have never discovered anything like this before."

The identity of the fish stumped experts until one claimed it to be a frogfish (which it looks nothing like). Another settled on a shortnose batfish.

Hispaniola

The island of Hispaniola lies between Cuba and Puerto Rico; it is the home of the nations of the Dominican Republic to the east and Haiti to the west. At 29,418 square miles, Hispaniola is the second-largest island in the West Indies. Christopher Columbus visited the island, and nothing's been the same since. The island has beaches (of course; it's an island), mountains, plains, and farms that produce corn, plantains, bananas, sweet potatoes, yams, rice, and cassava. Although on the same island, the Dominican Republic and Haiti are quite different, mostly a result of the mountain ranges that cut the island into the two nations. The mountains stop rainfall, plentiful in the Dominican Republic, resulting in a semiarid climate for Haiti. The Dominican Republic is a politically stable representative constitutional democracy, whereas Haiti is considered an authoritarian regime. Baseball is the most popular sport in the Dominican Republic; in Haiti, it's soccer. The Dominican Republic's economy is around 1,000 percent larger than its Haitian counterpart. Then there's the woman with backward feet.

La Ciguapa

The beautiful dark-haired woman is only seen at night. She wanders the forests, occasionally appearing in village kitchens in search of food. If unsated, the woman is known to lure men into the forest or into the mountains with a birdlike cheep only to devour their flesh and swallow their soul.

Sort of. The description changes in different areas of the Dominican Republic.

La Ciguapa is either a demonic entity/omen of impending doom/lovemaking-bloodsucking succubus, or a timid vegetarian who only kills if a man sees her in the forest. She will then hypnotize the unfortunate wanderer before it kills and eats him.

The brave who seek to hunt down la Ciguapa might find the demon difficult to track. Its feet are backward.

El Cuco

If the Caribbean has a boogeyman, it's el Cuco, a creature horror author Stephen King used as the villain in his 2018 novel *The Outsider.*

Depending on which country tells its story, el Cuco is either a human woman with the head of an alligator, or a human man with no face. Either way, it can alter its appearance from this form into something more familiar to its victim—and the victims are usually children.

With that, parents use the legend of el Cuco to keep their kids safely inside at night, and away from monsters that lurk in the dark; a Caribbean stranger danger.

El Cuco lives outside of towns in the forests and mountains of Hispaniola. When hungry, it departs its secluded den to wander into populated areas to look for troublesome children. It approaches them after shape-shifting into a safe figure and snatching them away, only to devour them in the darkness.

In the Dominican Republic, parents paint el Cuco with a more wicked brush. They tell their children if they misbehave or are disrespectful, el Cuco will enter the home and snatch the child straight from their beds.

So, kiddies, listen to Mom and Dad. You never know when the boogeyman will get you.

Zombies

Hollywood has used viruses, radioactivity, and fungi to turn people into zombies. That's not real life. But, *are* there zombies in real life?

In the case of Clairvius Narcisse, the answer is yes.

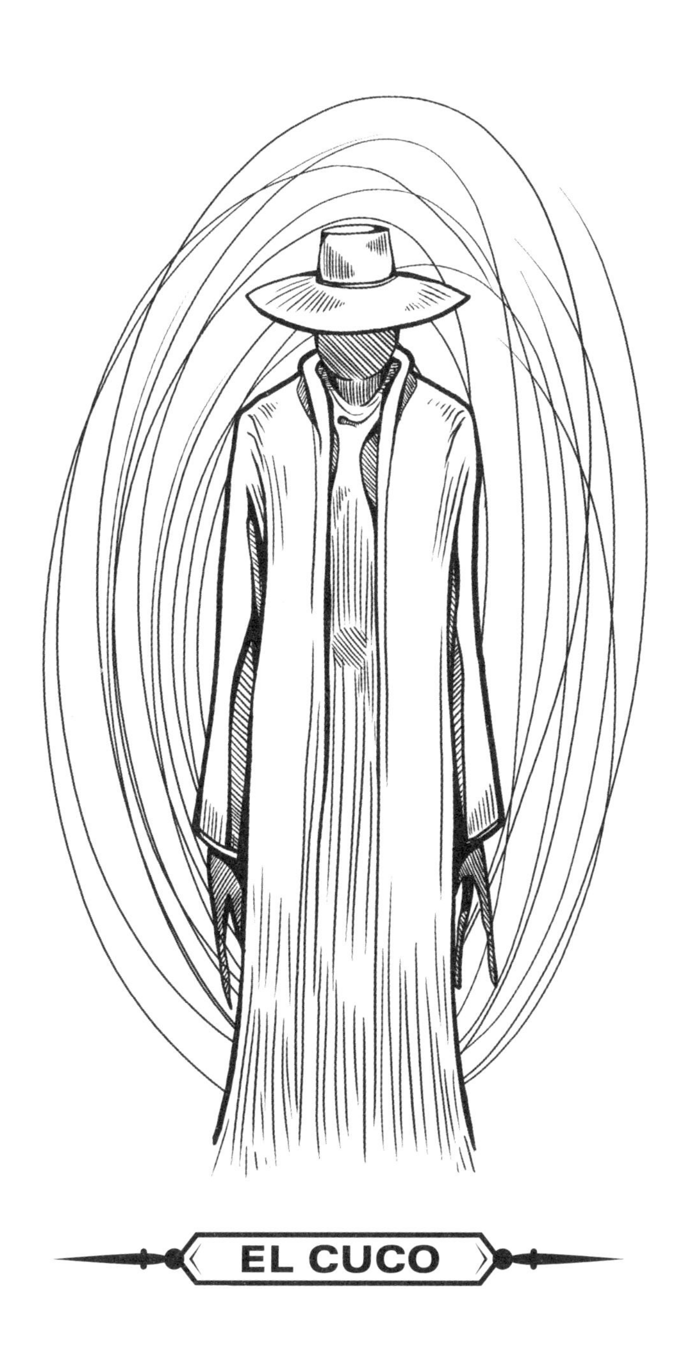
EL CUCO

He was born in L'Estère, Haiti, on January 2, 1922, and lived an apparently uneventful life until 1962 when he died three days after falling ill while chopping wood.

Dying is not the notable part. At least not the first time.

Eighteen years later, in 1980, a man wandered into L'Estère and said his name was Clairvius Narcisse. Locals went to fetch Angelina Narcisse, Clairvius's sister, who was, suffice it to say, a bit put out. The man convinced Angelina he was Clairvius when he disclosed information only the Narcisse family would know, including a nickname he called Angelina as a child.

So, if this was Clairvius Narcisse, what happened?

He claimed a Vodou priest drugged him, which made him sick, and later appear to be dead—although he was awake and aware during his burial, he simply couldn't move. Later, the priest dug him up, kept drugging him with a psychedelic concoction, and forced him into slavery on a plantation. This made sense to many people of the village because zombies have always been part of Haitian folklore.

Western science, of course, denies the existence of zombies, but Narcisse kept up the story of his drugging and enslavement until his second death in 1994. He didn't get up from that one.

Pontarof

The pontarof (also called a thébuch or tebuch) is a manta ray–shaped, dolphin-sized fish with a human head and huge bat ears.

Sexy, right?

It's also called the evil fish from its deadly nature. Lurking in the bottom of the ocean, river, or lake, this creature lies in wait for a child to enter the water, then it attacks. It never immediately kills the child; it plays with it first, often to the point of drowning them. Then, it disappears into the depths.

Jamaica

The third-largest island in the Caribbean (Cuba and Hispaniola have it beat), Jamaica has plenty to brag about. Resorts, rainforests, mountains, beaches, and, of course, the birthplace of reggae king Bob Marley. Is Jamaica known for great jerk food? Check. Great coffee? Check. Great rum? Check. Great beer? Check. What the hell am I still doing in the Midwest United States? I don't know. Jamaica was settled by the French, Spanish, and finally the British, at least until 1962 when it declared its independence, but remained a commonwealth of the UK. The island has a vast variety of ecosystems, including rainforests, alluvial forests, limestone forests, and dry, sandy spots with cacti. The island also has the Duppy.

Duppy

Another creature to come to the islands from Africa, the Duppy is known throughout the Caribbean, although the legend is strong in Jamaica. There, the Duppy is a spirit born from one of the two spirits Jamaicans believe a person possesses. A godly spirit that travels to heaven upon death, and an earthbound spirit that stays with the body.

The earthbound spirit may become a Duppy.

Bound to the body for three days after death, the Duppy appears in the guise of a shadow person. However, the Duppy can transform into the solid figure of a man or an animal. The behavior of the entity depends on the behavior of the person in life. A good person makes a helpful Duppy, a bad person makes a Duppy that can wreak havoc upon a person's life, even stooping to murder.

DUPPY

Duppies have been known to terrorize their widow, bankrupt their brother, and burn the house of a former business partner they believe had done them wrong in life. Duppies are unsavory sorts.

However, the living can take precautions against these entities. Before those first three days have passed, toss peas into the grave, or lay a cotton tree limb on the lid of the coffin. These methods will keep the Duppy from escaping the grave. If a Duppy is already loose, eating salt dissuades the Duppy from pursuing you, or wearing your clothes inside out may confuse them.

Happy Dupping.

Puerto Rico

Puerto Rico is an island of sand, sun, forests, and mountains. This land was settled by several groups of American Natives starting around 2000 BCE, colonized by Spain in 1493, then taken by the United States after the Spanish-American War in 1898. It is considered an unincorporated territory of the United States; its citizens have also been citizens of the US since 1917. Due to the island's beauty and climate, tourism is an important part of Puerto Rico's economy. Conspiracy theories abound (partly due to the US military presence on the island), including many UFO encounters, questions about the now destroyed Arecibo Observatory, and, oh, yeah, the Chupacabra.

Chupacabra

Madelyne Tolentino saw an animal around 4:00 p.m. the second week of August 1995 outside the house she shared with her mother in the Barrio Campo Rico neighborhood in Canóvanas, Puerto Rico.

She had no idea what kind of beast it was.

This bipedal creature stood about four feet tall with long arms tipped by three-fingered hands, and long, thin legs with feet "like a goose's foot." But it was the eyes that disturbed her most. They were enormous, slate gray, and bulbous—there were no whites. Parts of the creature were covered with hair; however, "pinkish-purple" burn marks spotted its body.

Tolentino may have thought she was hallucinating, but she wasn't the only person who saw it.

"I noticed that a vehicle was about to park right outside the house," she told Lucy Plá in a 1996 interview for *UFO Digest*. "It was then that I noticed that the fellow driving the car was frightened—his eyes opened wide and he started backing out."

A row of what Tolentino described as "feathers" connected by red skin ran down the beast's spine.

Tolentino screamed from inside the downstairs living room, and her mother came into the room and moved to go outside to get the animal, but it took off, not running, but hopping. She told Plá the thing's movements were "robot-like" and "slow, as if guided by remote control. Like a robot."

Her mother followed the monster, and a nearby young man tried to capture it.

"When he tried to grab the creature, it whipped out what I thought were feathers," she said in the interview. "He says that they stood straight up and that they were very long spines."

The creature bared its vampire-like fangs at the young man, and hopped off again, chased by Tolentino's mother, the young man, and a neighbor. It disappeared behind a church.

According to Tolentino, others saw the monster after dark, and its eyes glowed bright orange in the night.

CHUPACABRA

Tolentino saw the creature again, as did her husband. It smelled of battery acid.

"To my mind, it was nothing of this world," she told *UFO Digest.* "Look, it was neither animal nor human, you know?"

A number of locals saw the creature in the weeks that followed, including the minister of the church it disappeared behind. Then goats began to die, drained of blood with two puncture wounds on their necks.

This is the first report of the *Chupacabra* (goat sucker), a monster whose sightings have spread across Puerto Rico, the North American continent, Russia, and the Philippines. Its cultural presence extends from *The X Files* and *Unsolved Mysteries*, to music, comic books, movies, and the Disney TV show *Gravity Falls*.

But, is it real?

After the rash of sightings in Puerto Rico, the Chupacabra began appearing in Mexico and the United States, with some people claiming to have captured a live (and/or dead) Chupacabra. The creatures certainly look bizarre, although these entities were never four-foot-tall, bipedal, three-toed, or had feather-like spikes down their backs. After medical examination, each one turned out to be a dog, fox, coyote, or raccoon, each one with mange.

So, what did Tolentino see? According to Jonathan Jarry M.Sc. of Montreal's McGill University, almost anything.

"There are many conflicting accounts of exactly what this creature is supposed to look like," Jarry wrote in an article for McGill's Office for Science and Society. "Putting aside the mangey dogs of the mainland, descriptions of the Puerto Rican Chupacabra are all over the map. It has wings...or not. It has the ability to change colors...or not. It has a prominent tail...or no tail at all. Its ears are huge...or absent."

Referencing work by Benjamin Radford, at the time an author and columnist for *Skeptical Inquirer* and *LiveScience*, Jarry points Tolentino's description of the monster to that of the monster in the 1995 American science fiction film *Species*, starring Natasha Henstridge as a shape-shifting alien that has arrived on Earth to breed.

The movie was released in Puerto Rico in that year, and the image of Henstridge transforming into a bipedal alien with red eyes, spikes on the back, and long, thin fingers was already familiar to Tolentino.

"We rented a movie called *Species*," Tolentino told *UFO Digest*. "The movie begins here in Puerto Rico at the Arecibo Observatory. There's an experiment going on in the film. There's a girl in a glass box as a result of the experiment...What came out from inside the girl? It made my hair stand on end. It was a creature that looked like the Chupacabra, with the spines on its back and all. It even sticks something out of its mouth. Why did they make that movie? Why did they make it, and why in Puerto Rico?"

Was Tolentino influenced by the movie? Who knows. Even if she was, this doesn't explain the similar sightings by her mother, neighbors, pastor, and people around the world, nor does it explain the widespread attacks on medium and small farm animals in which all their blood is drained through two puncture holes.

Frogs

Not paranormal, and not cryptid, but Puerto Rico has a case of raining frogs.

The coqui, a light yellow or brown tree frog, is only about an inch long, but is loud enough to be named after the sound it makes. Although they usually live on the ground and in bushes, high humidity forces these frogs up into the trees of Yunque National Forest.

However, friendly neighborhood tarantulas, and other predators, have adapted to the frog's behavior and wait for them to climb. The frogs escape by jumping from the trees and raining back to earth.

St. Lucia

St. Lucia is a mountainous island nation in the south Caribbean. Famous for its beaches, resorts, reef diving, rainforests, cliffs, and… stop me if you've heard this before; it's a tropical island. Called the Helen of the West in comparison to the beauty of Helen of Troy, the Arawak Indians originally called St. Lucia Louanalao, which means "Island of the Iguanas," according to the Saint Lucia Tourism Authority. The island is also home to the Father of the Forest.

Papa Bois

Papa Bois, or Father Wood, is the protector of forests in St. Lucia, and the nearby island nations Trinidad and Tobago. African and French influences both went into naming Papa Bois, and both went into crafting various guises the protector takes.

He can appear as an old man in tattered clothing. The man, covered in hair, resembles a Greek satyr, with goat horns, and he walks on cloven hooves. He can also take the shape of an animal or a tree. Papa Bois protects animals by blowing on a horn made of bamboo whenever a hunter approaches. He is also a healer of the sick.

But beware, Papa Bois is vengeful in his treatment of unethical hunters, turning them into animals, or making them become lost in the forest.

PAPA BOIS

Trinidad and Tobago

The dual-island nation of Trinidad and Tobago lies 470 miles off the coast of Venezuela. Known for its music, beaches, rainforests, mangrove swamps, and the Asa Wright Nature Centre, Trinidad is not just gorgeous, it's also home to thriving industry. The smaller Tobago has white beaches and the hummingbird preserve, the Tobago Main Ridge Forest Reserve. The islands feature hiking, swimming, bird-watching...you know, the Caribbean usual. It's also haunted by the Douen.

Douen

The creatures are small, the size of children. They wander the night, their melon-sized heads and glowing eyes hidden beneath bulbous straw hats, crying to lure hunters and children into the forests, only to lose them in the darkness. The Douen are cursed with backward knees and feet, and always sport a hat to mask the fact they have no face. These entities are thought to be the souls of young people who haven't been baptized. Given that 63.2 percent of the country is Christian (mostly Roman Catholic), baptism is an important part of their faith.

Often armed with a whip of dried banana leaves, the wickedly playful Douen are fond of pranks and are known to throw objects at people and call to them, simply to get them to follow.

La Diablesse

Beware the Devil Woman.

La Diablesse, a legend of French and West African folklore, is as lustful as she is beautiful. On the night of a full moon, this gorgeous lady waits along seldom-used paths in the forests dressed in

DOUEN

large, puffy, colorful skirts and a hat, waiting for a man to appear and offer help.

Once a man sees her, she seduces him with dance and spell, charming him into following her, leading him up a mountain toward a cliff. If the man is fortunate, he sees what the skirt hides and escapes, because instead of feet, she has the hooves of a cow.

If the man is unfortunate, she takes his hand in her cold one and, as she transforms into the monster she is, demands he kiss her. Terrified, the man falls from the cliff to die at the rocky bottom.

Lagahoo

In a case of French influence, the Lagahoo is a shape-shifter, appearing as a man or woman during the day, but at night they change into a headless man carrying a wooden coffin. A potential victim can locate the presence of a Lagahoo by the sound of rattling chains. In other legends, the Lagahoo may change into a pig, a goat, or a wolf.

In some stories, the Lagahoo is referred to as the Devil's Businessman.

Mermaids

Disney put the image of the mermaid into the culture, but the behavior of a Disney mermaid and the mermaid of legend aren't exactly the same. The mer-being is a beautiful young person (woman or man) who appears human down to a lower half of a fish. Mermaids linked to these islands inhabit the waters north of Tobago and are known to grant wishes that are often coupled with a cursed trade.

Mermaids seek out interactions with humans, sometimes seducing them to remain in the water with the fish-person forever.

Chapter 4
Costa Rica

Costa Rica, sandwiched between Nicaragua and Panama, is the second-southernmost country in North America. This mountainous, rainforest-covered country is slightly smaller than West Virginia at 288 miles long and 170 miles wide. The Caribbean Sea graces the eastern beaches of Costa Rica, and the Pacific Ocean its west. Costa Rica is also explosive, with six active volcanos and sixty-one dormant ones; it has the twelfth most volcanos of any country on Earth. One-fourth of the country is protected jungles that team with various species of monkeys, brightly-colored tropical birds, and the crab-eating raccoon (not to be confused with eating crab Rangoon). Costa Rica's top exports are coffee, bananas, and pineapples. They keep their monsters to themselves.

El Cadejo

Opposite the Cadejo from Belize mythology, el Cadejo appears as a black chain-dragging dog, huge with blazing red eyes, the teeth of a jaguar, and goat's feet.

In a legend from the Costa Rican capital city San José's suburb Escazu, the Cadejo was once a young man who became drunk one night, and his father cursed him for his intoxication. The curse turned the man into a black dog doomed to wander the nights to terrify drunken people into giving up the drink.

Although the Cadejo is terrifying, it doesn't hurt anyone; it tries to change their lives.

La Cegua

Traveling through the mountains brings beautiful views and many dangers, both known and unknown. The unknown in Costa Rica is la Cegua.

Men—only men who are married or in an otherwise pledged relationship—who drive a vehicle or ride a horse along lonely mountain paths may encounter a beautiful woman. The woman always asks for a ride. If the driver is beguiled by her beauty and agrees, this woman's face melts into that of a horse skull, rotting meat hanging in shreds from the bone, its horse teeth jagged and filthy.

La Cegua's red eyes glare at the unfaithful man, and she kills them with a kiss of death.

Occasionally, a man may escape during the kiss, returning home with a bite mark on their face, branding them forever unfaithful.

Like many monsters in the southernmost reaches of North America, la Cegua is akin to the X'tabay of Belize.

La Llorona—Part 1

From the southern reaches of the United States to North America's southernmost country Panama, the Weeping Woman occupies a frightening place in Latin American mythology.

The legend begins with the Indigenous Bribris people of Costa Rica. They believed running water sounded like a crying woman and equated this to water spirits weeping for their lost children.

The far-reaching story of La Llorona is of a mother who throws her children into the river. Once she comes to her senses and realizes what she's done, she wanders the waterways in her grief, crying and searching for her children. She never realizes she's died and continues to look for her abandoned babies.

Begare

During the fourth voyage of Christopher Columbus to the New World, his men encountered a strange creature in what is now Costa Rica.

Columbus wrote in correspondence to the monarchy of Spain, one of his men injured the animal with an arrow. The animal "appeared to be an ape, except that it was much larger and had the face of a man."

Unable to approach the creature in its fit of pain and anger, they cut off one arm and one leg and threw it into a pen for a hog to eat. The pig, frightened of the animal, attempted to run away, but the creature attacked it with its tail, wrapping it around the pig's head and neck.

What was this animal? A type of simian like Columbus suggested? Geoffroy's spider monkey is the largest primate in Costa Rica at eighteen pounds. Although it has a prehensile tail like what

BEGARE

the explorer described, the animal simply isn't anywhere large enough to attack a pig.

German anthropologist Herbert Wendt considered the creature may be some sort of cat, although Columbus never referred to it as such.

Chapter 5
El Salvador

The Republic of El Salvador is 8,124 square miles of plains, mountains, jungles, and beaches. It's the only country in the Central American portion of North America that doesn't have a Caribbean beach, although its entire southern border is 190 miles of Pacific shoreline; its other borders touch on Honduras and Guatemala. The smallest of the North American countries, El Salvador isn't small on history. It's been home to the Olmec, Maya, Toltec, and the agrarian Pipil peoples (descendants of the Aztecs), until the Spanish invaded in 1524. Unlike most of the world that goes through four seasons, the country has two—wet and dry. The wet season sees nearly sixty inches of precipitation (more than a foot per month between June and September); the dry season between December and March is called dry for a reason. It almost never rains. Like all other equatorial North American countries, El Salvador has its share of monsters. Not listed below are creatures such as el Duende, Siguabana, el Cadejo, and la Llorona.

Countries with similar cultures have similar monsters. But, don't worry, El Salvador has its own legends, like el Caballero Negro.

El Caballero Negro

The Devil rides in El Salvador.

El Caballero Negro, the Black Knight (or Horseman, depending on the translation), regally dressed as a gentleman, rides his dead-black horse over rural roads and city streets, a wide-brimmed hat obscuring his identity. This figure stops at crossroads and propositions the sad and desperate. Like a magician, the black gentleman produces a coin and offers it to the person. If the person refuses it, he bestows blessings of riches and second sight to them for seven years and will then return to take them to Hell. If the person accepts the coin, he has sold his soul to the Black Knight and will spend his life tortured by the fear of the Devil chasing him down.

Some legends have el Caballero Negro promising the passersby money, love, happiness, or whatever the doomed wish the most. The cost in this case? A boy will be taken from the family for seven generations. After seven generations, the home in the mountains or an entire village will reek of sulfur. A sure sign the Devil had come to visit.

So, if you're traveling around El Salvador at dusk and a dark, handsome man on an equally dark horse stops to say hello, just keep walking. You may become the victim of seven years' good luck (and seven generations of bad).

El Tabudo

Given the fact that El Salvador has 190 miles of access to the Pacific Ocean, 360 rivers, and six notable lakes (four of any real size),

EL CABALLERO NEGRO

fish isn't only one of the country's greatest exports, they're a way of life. So, there has to be a legend of a fisherman.

El Tabudo was a fisherman taken by the ocean. The villagers knew he had been drowned, but the man came back alive, although his knees had become large, and knobby. In some stories, the fisherman has returned as a, um, a manerfish? El Tabudo has human legs (with huge knees) and a body like a fish—sometimes.

This entity lurks near the water—of lakes, rivers, or the ocean—and greets those who walk by as seemingly an old fisherman. Once he reels them in with his huh-yuck charm, el Tabudo coaxes them into the water and transforms before turning the men into fish, and the women into mermaids.

At Coatepeque Lake, the story is slightly different. The man wasn't just a fisherman, he was wealthy. When he went out on the lake in a boat, the water goddess kidnapped him. Soon, the man, his knees huge, and his mouth that of a fish, gave his riches to his servants before retiring to the waters of Coatepeque Lake forever.

El Padre sin Cabeza

The Father was a player.

During the Spanish colonial era of El Salvador in the 1500s, a handsome priest (some legends have him serving in San Salvador, others in nearby León, Nicaragua) became the obsession of many women in his parish. Although he never forsook his vows of celibacy, he became tempted by a beautiful woman who, in confession, professed her love for him. He asked to be transferred, but was denied. Eventually, he gave in to the woman's advances, and the two began an affair.

Rumors spread about the couple, and the locals censured the priest for disobeying church doctrines. Embarrassed, the priest fled into the country, only to be later captured and brought back to town for a trial. A judge sentenced the priest to death by beheading.

Before he died, the priest cursed the town, saying his soul will wander the town and nearby roads, headless, until all its inhabitants renounce their own wrongdoings.

An encounter with el Padre sin Cabeza (the Headless Father) begins with a pale glow, and the smell of incense. Although this headless figure invokes terror in those who see him, the father simply forgives the wrongdoers, and continues through the night seeking his own forgiveness.

La Cuyancúa

An ugly beast lurks in Izalco, a city in the state of Sonsonate.

La Cuyancúa, a creature from Maya mythology—called the Messenger of the Rains—has no equal in strangeness. It appears as a pig with the lower half of a snake, and its front legs (its only legs) are like human hands. La Cuyancúa comes before a storm, squawking in the night as it pulls its snake body along the ground with its human hands.

Some legends say if la Cuyancúa lies upon parched land, a spring of drinking water forms there. When it is hungry, it hides in irrigation ditches and awaits its prey. Those who encounter it are struck speechless with terror. For fun, it sneaks through streams and frightens women washing their laundry.

Not cool, man. Not cool.

Cipitío

Cipitío is a mischievous boy with a swollen belly and an enormous hat. This son of the once beautiful Siguanaba (see la Siguanaba in the Bay Islands chapter) suffered a similar fate as his mother who was cursed by the Aztec rain and fertility god Tlaloc after betraying the sun god by having sex with the Devil and becoming pregnant. Siguanaba became cursed to have the face of a horse and to murder those men who betray their wives. Tlaloc cursed Cipitío to never grow up and always appear as a boy of ten.

Like many other creatures of the region, Cipitío's feet face backward, so he is hard for someone to track. Making finding Cipitío even more difficult is his penchant to transport himself whenever he wills it.

Although difficult to find, Cipitío eats ashes and can be found near dead fires or furnaces. He also likes to hide near streams and waits for pretty girls to come to the water to bathe. Then he catcalls and whistles at them, throwing flowers and laughing. However, these women cannot see him. The only ones who can see Cipitío are children, like himself.

Chapter 6
Greenland

Greenland is, if you look at the country, a bit of a lie. It's not green at all. Farming (a really quite green endeavor) is relegated mainly to sheep production, and sheep, the last time I checked, aren't especially green. Greenland does produce crops like potatoes and turnips, but the important parts of these plants (namely the potatoes and turnips) have the good sense to stay underground. What Greenland does have a lot of is snow, ice, and glaciers, which are, again, not green. What green it does have is a forest—one forest—that stretches a little more than eight and a half miles. Eight and a half. Eight and a half miles of greenery in a white country called Greenland. We can blame Vikings for the misnomer (though not to their faces. They're a grumpy bunch). Erik the Red "discovered" Greenland in 983 AD (and by "discovered" I mean he didn't count the Inuit people who already lived there as having any claim on the land at all). By naming the island Greenland (the largest non-continent island on the planet), he hoped "green" would attracted people to settle there. Today, Greenland is a country in the Kingdom of Denmark, granted home rule by Denmark in 1979. Eighty-eight percent of the population are Inuit, the rest mostly Danish. This country is 833,981 square miles with a population of 56,653. The capital city, Nuuk, has a population of 18,000. Outside the cities, towns, villages and settlements, there are no roads or railroads. People travel by dogsled, snowmobiles, helicopters, airplanes, boats, or they hoof it (which isn't recommended because of all the polar bears). The major industry is fishing. The Northeast Greenland National Park is the largest national park in the world at 386,102 square miles, and at 694,984 square miles, Greenland's ice sheet is second in size only to Antarctica.

The place is big so, of course, it has its own monsters.

Erlaveersiniooq

Greenland monsters are humanlike, and ravenous. The Erlaveersiniooq is no exception.

Much like legends in other countries, the Erlaveersiniooq is a warning not to go wandering off. This creature is a woman who lives in a lonely shack in the wilderness. If someone nears her hovel, she attempts to bring them inside with the promise of a warm fire, and a peaceful rest. If they accept the invitation and enter her hut, she pounces, slicing their belly open with a knife.

Then, she cooks and devours their viscera. No word on if she eats their liver with a nice chianti.

How to defeat an Erlaveersiniooq? Eat your own intestines and liver before she can. Yum, yum.

IKUSIK

Ikusik

Ikusik is the Greenland Inuit word for "elbow." It's another humanlike entity that enjoys eating living people.

Erlaveersiniooq? Ikusik? You two, get a room.

The Ikusik is a human corpse that only attacks if a person has desecrated a grave. It is called elbow because its arms have rotted off up to the elbows, and it uses its elbows to propel toward its prey, dragging its useless legs behind it. Even though the Ikusik is traveling on its elbows, it can move fast enough to catch anyone running from it.

Then it eats the screaming victim alive.

Qivittoq

If a person (usually a criminal) is banished from a village and tossed into the deathly cold, they become a Qivittoq when they perish; a ghostly entity that wanders the frozen wastes of Greenland, shape-shifting into animals if they so choose.

This monster, evil by nature, inhabits the frozen hills, preying on anyone who wanders too close to its territory. Hungry and cold, the Qivittoq—like the Erlaveersiniooq and Ikusik—eats those people it captures.

It is not the people of Greenland only who can become a Qivittoq; the land, snow, air, and sea—everything that possesses a soul—judge those who have harmed nature, and treat them accordingly. So stories of whalers who have perished upon the shores of Greenland involve these beasts becoming Qivittoq when they die alone in the cold.

Satsuma Arnaa

In the native language of Greenland, *Satsuma Arnaa* means "Mother of the Sea." And, just like your human mother, don't make her angry, or you will be punished.

The Inuit people of Greenland survive on the protein of reindeer, walrus, and whales (the Inuit are exempt from the international whaling ban). They also fish for Atlantic salmon, cod, Arctic char, and redfish. However, Satsuma Arana does not tolerate waste from the sea.

Respect the animal. Honor the animal. Use every part of the animal you have killed. If not, Satsuma Arnaa's long hair stretches out and tangles around the creatures of the sea, keeping them safe from the greedy, and those ill-suited to care for her children. Those families get no food, no hides, no bones.

So, when the Indigenous peoples of Greenland have an unsuccessful hunt, or a poor catch, they are being punished by Satsuma Arana. In comes the shaman who goes to Satsuma Arnaa and begs for forgiveness while he combs her hair, telling her his people will be better stewards of her children. This pleases the mother of the sea, and her hair releases the animals for the Inuits to harvest again.

Tupilak

Shamans aren't only known for calming the wrath of Satsuma Arnaa—they can also create flesh golems.

Tupilaks (also spelled *Tupilaq*) are the spirits of ancestors shamans trap in stolen corpses (sometimes of children), modified with parts of animals like caribou and walrus, then coupled with an item that belongs to a person who wronged the shaman. The Tupilak then comes alive, and this monster becomes a tool to destroy the enemies of the shaman.

Shamans can rid themselves of the Tupilak by discarding it into the ocean.

Chapter 7
Guatemala

The Maya K'iché people, who controlled Guatemala from the thirteenth century until the Spanish conquered it in 1524, named the land Cuauhtēmallān, Place of Many Trees. Along with the rainforest that covers nearly 35 percent of the country (it covered 50 percent as recently as 1950), Guatemala has mountains, thirty-seven volcanos (only three are active), and farmland that grows sugarcane, coffee, bananas, and cotton. The hot, humid country is bordered by the Caribbean Sea, the Pacific Ocean, Mexico, Belize, Honduras, and El Salvador. Named the fifth of thirty-six Biodiversity Hotspots on Earth by the World Conservation Union, Guatemala is home to monkeys, caimans, peccaries, ocelots, and, of course, monsters.

Quetzalcoatl

This creature has a long history.

Starting in around 100 BCE as the god of wind, farming, inventor of calendars, inventor of books, patron of priests...wow. In the mythology of the Maya and Aztecs, Quetzalcoatl did it all.

The image of the god began appearing in Central American art in the form of a snake with feathers, sometimes depicted alongside a conch shell (blow a conch shell, god of wind. Makes sense). Thirteen hundred years later, Quetzalcoatl was shown as a man, dressed like a Mardi Gras parade: tall Pope-ish hat, a red duck mask, and conch shell jewelry. But his story remained the same.

Although a more than one-thousand-year-old story can change, according to some Aztec and Maya myths, Quetzalcoatl was born to creator gods Ometeotl and Chimalma, a brother to Xipe, Totec, Tezcatlipoca, and Huitzilopochtli. Then things get weird. Legend has it Chimalma married Mixcoatl after he shot arrows at her, and she caught them (*Chimalma* means "shield hand"). Pregnancy, however, evaded her. It was only after praying to Quetzalcoatl, and swallowing a jade stone, did she become pregnant, and later give birth to the five brothers—including Quetzalcoatl.

Quetzalcoatl's place in the creation mythos came when his parents gave the children a part in assembling the universe. He created fire, corn, then the Adam and Eve of Maya and Aztec mythology.

Although the gods disappeared into the netherworld, the Aztecs believed Quetzalcoatl returned in 1519 in the guise of the Spanish conquistador Cortes. Due to some similarities to Christianity, at the time the Catholic Church considered Quetzalcoatl to either be the apostle St. Thomas, or Christ himself.

All this mythology, could Quetzalcoatl have been real?

Possibly, the Feathered Serpent is reflected in the green, blue, and red resplendent quetzal bird (the national bird of Guatemala). During mating season, males grow flashy, brightly colored feathers resembling a bride's train. The bird, thought by some to be a spirit guide, is a sacred, treasured creature of Mesoamerica; so treasured, the money of the country is called the quetzal.

However, in the Totonicapán Department (province), Quetzalcoatl means something else. It's not a god, it's not a bird, it's a legendary snake—one to be feared.

A Quetzalcoatl is a deadly mythological Guatemalan water snake identified by the green feathers around its head, the blue feathers down its back and tail, and the red feathers that decorate its underside. Its beauty is short lived because this flamboyant monster's venomous bite not only kills its prey, it kills itself.

Yeah. That's it. The lesson? When in Guatemala, avoid feathery snakes; they're bad.

El Sombrerón

The small man with the large black hat is irresistible.

The tale of el Sombrerón is from the seventeenth century; in some regions the woman who falls in love with the large-hatted man is named Celina, but in Guatemala, she is Susana of Antigua.

As the story goes, one night when Susana sat on her balcony brushing her long brown hair, she heard a ruckus coming down the path. Soon, a man in a black hat emerged from the darkness leading four mules. He stops beneath her balcony and begins to sing. Susana falls in love with the man. When her parents (wiser than her) rush out to stop the man, he vanishes, and Susana's heart breaks. She couldn't go to sleep that night, and when she woke, she couldn't eat, and began to slowly die while she waited for the mysterious man to return to her.

Eventually, he did.

As she lies sleepless in bed, she again hears the song, and finds the man on her balcony. He enters her room and comforts her. Susana's parents soon come into her room and find the man braiding their daughter's hair. El Sombrerón escapes from Susana's angry parents, and, afraid the man was casting a spell on their daughter, they cut off her unfinished braid and take it to be blessed at their local church.

After the blessing, Susana returned to normal, and never heard the man's song, or saw him, again.

Susana, of course, got off easy. Other legends say el Sombrerón is really the Devil and captures the souls of his victims. Once he finishes the braid, the women are tied to him forever.

This is similar to a Maya story where the woman becomes pregnant and her offspring wears the hat.

Whatever the case, people throughout the centuries have reported running into the charming el Sombrerón when the sun sets and the moon is high.

El Huay Chivo—Part 1

El Huay Chivo, the sorcerer goat, often appears as half-man/half-goat, although it can also transform itself into part-dog, deer, or horse. The story of el Huay Chivo comes from the ancient Maya.

This wizard, who can only change his shape at night, walks on two legs. It has black fur and red, glowing eyes—it's four feet tall. At this point, this magician begins to sound a bit like Puerto Rico's Chupacabra, and it has also been accused of killing goats and chickens. Unlike the Chupacabra, some stories insist el Huay Chivo must remove its head before it can transform into a half-animal.

El Huay Chivo is also said to be accompanied by a gust of wind and a vile odor. Which brings us to the Maya god Cizin, the stinking one, the god of death, earthquakes, and, um, flatulence.

The story of the Huay Chivo began when a young sorcerer fell in love with a goat farmer's daughter. Turned away by the farmer, the sorcerer made a deal with the Devil (or, in this case, Cizin) to become a goat so he could still be close to his love. As such deals go, he became half-goat, although his soul is still owned by Cizin.

Although a goat-man may seem outrageous, accounts of the Greek satyr—and sightings of such a creature in Fort Worth, Texas; Point Pleasant, West Virginia; and Pope Lick, Kentucky—may mean el Huay Chivo is alive and well and roaming the earth looking for his lost love.

Ix-hunpedzkin

A relative to the Gila monster of the American Southwest, the Ix-hunpedzkin is a species of venomous beaded lizard native to Guatemala. At between three and four feet long, these black, gray, and pink lizards are beautiful and intimidating; however, it's not just their bite a person should worry about.

Here's where the Ix-hunpedzkin becomes legend.

Although the lizard possesses venom glands in its mouth, its whole body is toxic. If one of these creatures simply touches someone's clothes, it leaves enough toxin to kill an adult human.

But wait, there's more. It's not sold in any store.

IX-HUNPEDZKIN

If an Ix-hunpedzkin so much as bites the shadow of a person's head, it will inflict an agonizing headache that, if not treated with the leaves of the Tillandsia plant, will eventually kill its victim.

El Sisemite

A hairy figure dwells in the deep, lush rainforests of Guatemala, a cousin to a similar horror, the Sismite, from nearby Belize. This bipedal creature, tall as a young tree, resembles a man as much as it resembles a monkey; it is known as el Sisemite, the Maya Bigfoot.

This monster, much like the Douen and Jumbee, has backward feet that make it hard for a hunter to follow it. But even if a hunter is successful in tracking this elusive creature, it won't be able to bring it down. A Sisemite's hair is so long and thick, bullets cannot penetrate it.

El Sisemite is smart, but not smart enough. It envies man's ability to make fire and is known to eat the ashes and charcoal from dead campfires, even stacking sticks in a semblance of a campfire, although it can never light it. It is also jealous of man's ability to speak, occasionally stealing children to teach it how, although el Sisemite has never accomplished the feat.

Death follows any man who sees the Maya Bigfoot, usually by el Sisemite itself. However, women are granted long life. Unfortunately, that long life may be spent as captive in the cave of el Sisemite.

Chapter 8
Republic of Honduras

Bordering Guatemala, El Salvador, and Nicaragua, the Republic of Honduras also has a 430-mile Caribbean Sea coastline and a minor Pacific coastline that mostly consists of mangrove forests and swamps. This country, about the size of Tennessee, is made up of mountains, subtropical lowlands, and hides ancient Maya sites in its tropical rainforests. Around 30 percent of Honduras is set aside for agriculture, but nearly half of the country is covered in forest. Once part of the Spanish Empire, Honduras gained its independence in 1821, only to become part of the Mexican Empire two years later, then became a member of the United Provinces of Central America until it ended in 1838. Honduras is famous for coffee, sugarcane, textiles, and a creature affectionately called the Tongue Eater.

EL COMELENGUAS

El Comelenguas

The story starts with cattle—and birds. Big, big, birds.

In the 1950s, farmers near Nacaome (a city near Honduras's southern border and the Bay of San Lorenzo) saw an enormous animal they initially identified as a bird flying over their grazing land. The next day, farmers discovered some of their cattle had been killed.

One farmer said he saw the death occur; a monstrous creature swooped from the sky and grappled a cow with its long, serpentlike tail. The beast squeezed the cow until it stopped breathing, then ripped out the bovine's tongue.

From that day on, the locals called this monster el Comelenguas, the Tongue Eater.

As the occurrences increased, more witnesses claimed to have seen the beast, a giant bird with a long beak, huge wings, and a long, muscular tail with feathers. The creature would remove its victims' tongues by tearing them out with its strong beak.

Explanations for the Tongue Eater have ranged from a griffin, to a Thunderbird, to a pterodactyl, to a flying snake, to a huge bat.

One-Eyed Giants

The Indigenous Miskito peoples of Honduras and Nicaragua have a legend of forest giants with one eye. These cyclops-like monsters also have a head like a dog and a mouth where a human navel should be.

And, of course, they're cannibals. I mean, why not?

Legends of giants—especially cannibal giants—aren't strangers to North America. Stories of giants (usually red-haired giants) stretch from Mexico to Lovelock Cave in Nevada, Catalina Island off the coast of Southern California, to Illinois, and up into

Canada. These giants are all cruel, hungry, and slaughter humans, sometimes for the heck of it.

Pfft. Giants. Amirite?

La Taconuda

Some legends from Mexico through the Caribbean and Latin America have one thing in common—a beautiful angry woman on a lonely road. In Honduras and Nicaragua, it's la Taconuda.

This beautiful woman is around seven feet tall and dressed for a party. She was the daughter of a wealthy Spanish landowner named Sanchez. When he died, she inherited everything her father had owned, and men flocked to marry her. She was less than impressed with all of them.

Story has it that when she refused to pick a husband, two potential suitors broke into her house and slashed her face with broken glass. Although she lived through the attack, her beauty was gone.

Seeking revenge, she called upon the Devil to erase the damage to her face in exchange for her delivering him a soul per week. Although long dead, la Taconuda still roams quiet roads, looking for the next soul to sacrifice, and, although her figure is still beautiful, her victims find the face of a skeleton grinning at them before it takes their life.

Chapter 9
Republic of Nicaragua

Nestled between the Caribbean Sea to the east and the Pacific Ocean to the west, the Republic of Nicaragua (a co-founder of the United Nations) is a country rich in history and diverse life. Jungles, beaches, and lakes cover this nation, which is slightly bigger than New York State. Nicaragua—known as the Land of Fire and Water—is home to the second-largest rainforest in the Americas (unlike New York), the Mosquito Coast (named after the Indigenous Miskito Nation), and nineteen volcanos (extinct and active). Humans have lived in the country since 12,000 BCE, including (much later than 12,000 BCE) the Maya and Aztec empires. Animal life includes quetzals, goldfinches, eagles, macaws, anteaters, tapirs, monkeys, and, off the Pacific coast, the Nicaragua shark. There's also an enormous worm called the Minhocão.

Minhocão

The creature is big. Huge. Gigantic. Open your thesaurus, folks, because the Minhocão is in the house. Not exactly in the house. Maybe under the house, or through the house, the size of the house. It's big.

From Nicaragua to Brazil, this monstrous worm with horns and impenetrable black scales grows upward to a reported 150 feet long. It travels underground through jungle regions, surfacing for food and wanton destruction. The Minhocão has a piglike snout; its mouth is surrounded by tentacles. It eats large animals, such as horses, deer, and cattle, surprising prey by erupting from underground. Monsters from the Kevin Bacon movie *Tremors* (1990) were based on the Minhocão and the Olgoi-Khorkhoi, better known as the Mongolian deathworm.

One menace of the creature, apart from all the eating, is the destruction left in its wake. Homes, businesses, bridges, and roads all collapse due to ground weakened by the Minhocão's tunnels. Depending on the age and size of the monster, the tunnels can be anywhere from three to ten feet wide, and, since the Minhocão is most active during the wet season (May to October where around fifty-six inches of rain fall), its tunnels often become underground rivers. The worms, however, are said not to dwell underground; they are believed to live in large bodies of water.

Mentions of the creature stretch back to the 1800s.

Science fiction author Jules Verne (*Journey to the Center of the Earth, 20,000 Leagues Under the Sea*) mentions the Minhocão in his 1881 adventure novel *Eight Hundred Leagues on the Amazon*. Naturalists such as Benjamin G. Ferris, Augustin François César Prouvençal de Saint-Hilaire, and University of Edinburgh Professor

MINHOCÃO

Andrew Wilson interviewed witnesses of the worm, and not only found the stories similar, they considered them plausible.

If the Minhocão is out there, what is it?

Giant snakes are considered a possible source of the Minhocão legend, although the green anaconda of the Amazon Basin can reach up to thirty feet (just a tad shy of the Minhocão's 150 feet). The three largest snakes in Nicaragua are the Central American boa at eight feet, the puffing snake at nine, and the largest, the terciopelo, at fifteen.

Caecilians are also a candidate. These nearly blind, legless, snake-like amphibians live hidden underground, usually in wetlands, and they tunnel well. They are virtually unknown to most people. However, the longest caecilians only reach up to five feet in length.

Nineteenth-century naturalist Auguste de Saint Hilaire suggested the Minhocão may be a species of giant lungfish. The lungfish is a freshwater fish that can breathe air and lives in riverbeds. But, like the caecilians, it is much too small to be considered the source of the legend.

Living fossil? Unknown monster? Unless one is captured, the Minhocão is simply a story.

Other countries reporting sightings of the Minhocão include Uruguay and Bolivia.

Li Lamya

Nicaragua is home to a variety of felids, like jaguars, ocelots, cougars, margays, jaguarundis, and Li Lamya—the water tiger.

Li Lamya, also known as Was Nawahni by the Mayagna people of the Caribbean coast, is described as an aquatic big cat that sports webbed paws, like those of an otter. By its name, the beast is quite at home in the water, living around rocks and preying on any livestock, deer, or people who venture into its domain. However, it becomes awkward when it ventures onto land.

Author Eduard Conzemius wrote in his travel book *Ethnographical Survey of the Miskito and Sumu Indians of Honduras and Nicaragua* (1932) that Li Lamya are a generally black manatee-like creature, although the color of their coats may vary from black, to brown, to tawny. And they're tough to catch. The coastal Indigenous tribes, the Mayagna and Miskito, have failed at every attempt to capture or kill Li Lamya.

According to Natasha-Kim Ferenczi in her 2018 thesis, *What if There Is a Cure Somewhere in the Jungle? Seeking and Plant Medicine Becomings*, the water tiger is still around, and still dangerous. During floods in 1988, a water tiger killed "about ten Indigenous people."

Whether the Li Lamya exists or not, it certainly is a bad kitty.

La Carreta Nagua

The wagon only rolls by after midnight, and before dawn. Chains drag the road as the old cart comes round; the clanks drawing terror through the souls of those still awake to hear it. As the ramshackle cart nears, witnesses first see the oxen pulling it, their hide pulled tight over emaciated ribs. These unfortunate people hope to never see the cart driver—a skeletal being shrouded in white—because he (sometimes she) is the Devil.

The wagon never turns, it only moves straight down the road, vanishing at a corner and reappearing on the next street. People cowering from the vision of la Carreta Nagua pray the cart never stops, because when it does, someone dies.

Academic types (who are never any fun) attribute this mythology on the Indigenous tribes' fear of the conquering Spaniards who poured through the area kidnapping and pillaging. The story of la Carreta Nagua quickly became a bedtime story aimed to keep children in the house at night.

Glyptodont

Up until roughly eleven thousand years ago, the armadillo was terrifying.

The Miocene ancestor of the armadillo, the Glyptodont, was nearly the size and shape of a Volkswagen Beetle. Not that Volkswagen Beetles are frightening, but Beetles don't have bony armor and a heavy, spiked, stegosaurus-like tail.

Hmm. Come to think of it, if they did, I'd own one right now.

The Glyptodont's bulbous shell was about five feet long. It was omnivorous, eating plants, insects, and leftovers, you know, carrion. Being a living tank, the only predators that attacked the Glyptodont were giant bears, teratorns, saber-toothed cats, and other carnivorous megafauna.

Like with other ice age creatures, people have reported seeing the Glyptodont in modern times. However, much like academics, hoaxers ruin the fun for everyone.

A giant extinct armadillo was reported captured in 1967—even reported in the October edition of *Science News*. It was six and a half feet long and weighed more than two thousand pounds. Scientists speculated the Glyptodont may be extant. Unfortunately, a specimen was never produced, and the report was labeled a hoax.

Chapter 10
Panama

Panama is a famous isthmus. Say it out loud, "Isthmus." Fun, right? An isthmus is a thin strip of land (Panama is thirty-seven miles wide at its thinnest) that connects two larger pieces of land (in this case North and South America). What makes it famous is no doubt being the birthplace of Juan Williams, Fox News journalist...just kidding. It's the man-made Panama Canal that connects the Atlantic Ocean to the Pacific Ocean. The Panama Canal is a vital shipping route that cuts the trip from Point A to Point B by around twenty-two days and eight thousand nautical miles. More than sixteen islands are located off the Panamanian coasts. This S-shaped country is filled with mountains, semideciduous tropical moist forests, rivers, beaches, and bays. It's also home to la Tulevieja.

La Tulevieja

The beautiful woman sits by the water, her large round breasts bare. When she was young, she fell in love with a man in her village and became pregnant. They weren't married, and she couldn't bear the shame, so she fled into the forest. After she gave birth to a boy, she tossed it into a river and returned to the village, not shamed, but saddened deeply. Angry at what the woman had done, God punished her by turning her into a monster, doomed to wander the waterways of Panama forever, looking for lost children.

This is only the beginning of the legend of la Tulevieja. As she sits, she waits for men to pass by. If a man is entranced by her milk-filled breasts, she offers them to him. If he suckles on them, the beautiful woman transforms into a beast that is half-woman/half-hawk, her long lush hair now ratty and tangled, her hands claws, and her feet are backward (again with the backward feet?). After inducing terror, the lusty Tulevieja slices up the man with her razor-sharp claws.

However, la Tulevieja isn't entirely evil. If she finds lost children, she will feed them her milk.

You can keep la Tulevieja at bay with prayer, or, if you hear a woman crying near a river in the forest, it's best to simply leave her be.

La Silampa

When the Mother of Night comes, she comes on silken air. If a fog rolls off the mountains on a cold night, the sheet of white mist may be la Silampa.

The legend begins as many such legends do begin, with love. When a Native woman fell for the greatest warrior of her peoples—a forbidden love, since she was a princess—the couple had to flee. The king sent warriors out to bring back his daughter and kill the man. During the pursuit, she begs the night spirits to defeat the warriors, but this is too late, and her father kills her lover. The night spirits enter the woman, transforming her into the Mother of Night. She has since wandered the darkness in the guise of a fog, searching for men alone in the night, only to envelop them in her sheetlike embrace and devour them, leaving only their bones.

Las Brujas

Witches get stitches. No. Wait. Witches *give* stitches.

Las Brujas are women in league with Satan, selling their souls to el Diablo to possess magical powers. How do they keep those powers? By a big nourishing bowl of Sancocho de Gallina? No. They suck the souls from unbaptized babies through their belly buttons.

A person can tell if a witch is nearby because someone has braided their horse's tails. Las Brujas are also fond of whistling. They can appear as animals and are quite fond of transforming into a deer. Las Brujas can also fly.

To avoid a witch, turn your clothing inside out (much like the 1990s hip-hop group Kris Kross, "'Cause inside out is wiggity-wiggity-wiggity-wack").

Las Brujas is also, of course, a brand of Panamanian coffee. Bottoms up.

El Monstro

Sharks abound in the waters of Isla de Malpelo, a tiny Pacific island (one mile long, 2,100 feet wide) about 373 miles south of Panama. Deemed one of the best places for shark divers in the world,

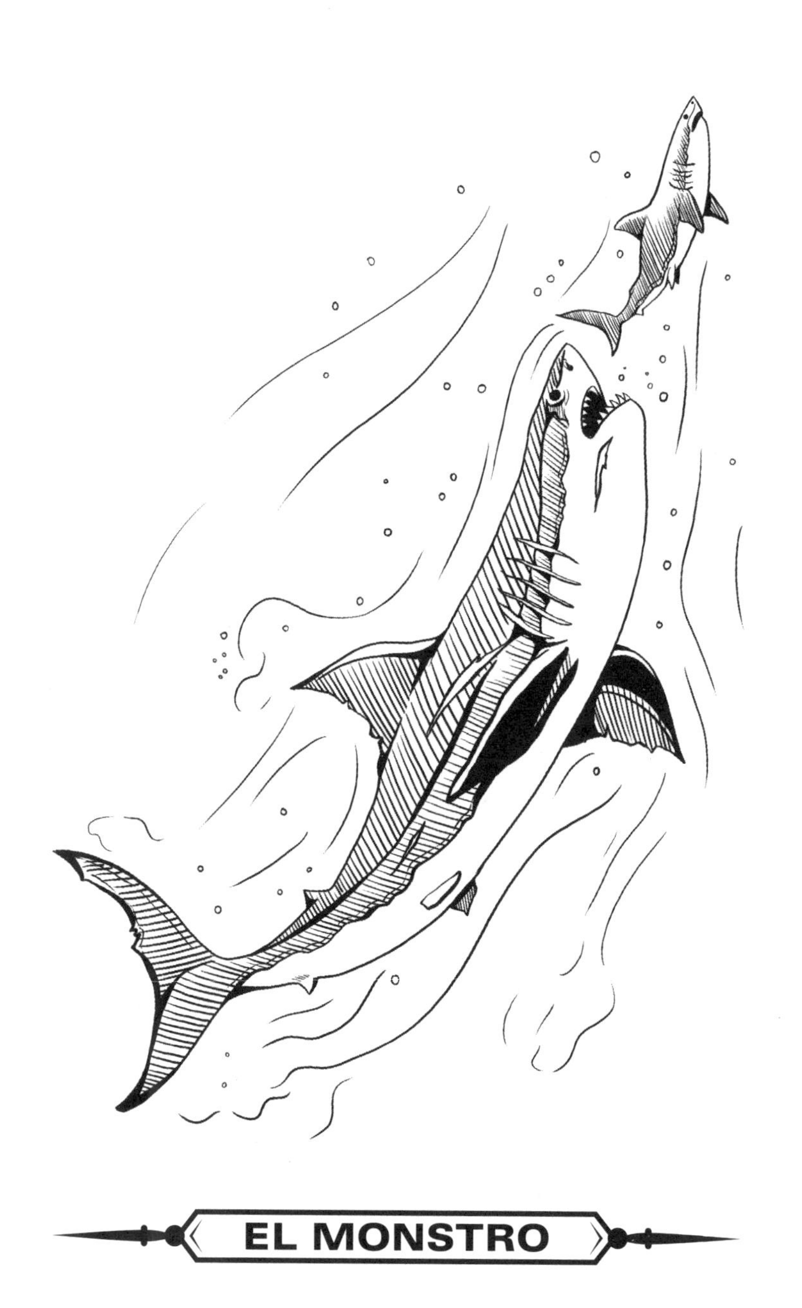
EL MONSTRO

according to the Project Malpelo website, the island "is one of the last remaining marine ecosystems on the planet where divers can still encounter a thousand sharks on a single dive." The island itself is uninhabited, composed mostly of rock, the highest of its three peaks, Cerro de la Mona (Mona Hill), reaches only 980 feet.

The island's waters are also the home of a beast.

The Malpelo Monster, or el Monstro, was reported by biologist Sandra Bessudo who, while diving, encountered a mysterious species of giant shark. The creature is a fifteen-foot-long sharklike fish with abnormally large eyes and a dorsal fin directly over its pectoral fins, as opposed to behind them, which is seen in other sharks. Bessudo encountered the monster fish in a dive below 160 feet.

People have speculated el Monstro is one of the four species of sand tiger sharks, although none of the species match the length, eye size, or fin placement of the Malpelo Monster.

The Panama Creature

In September 2009, teenagers exploring a cave in Cerro Azul, Panama, claimed to have encountered a monster. Terrified, they hit this beast with sticks and pelted it with rocks until it stopped moving. After tossing the body into a pool, they sent pictures of it to a local television station.

The story went viral and photos of this pale, bloated creature circled the globe, drawing comparisons to a similar find in July 2008 in New York State dubbed the Montauk Monster (which later was found to be a hairless, waterlogged raccoon).

People speculated it was an extraterrestrial creature, or something similar to Gollum from *The Lord of the Rings*, or a dead Chupacabra, or…

Oh, hell. It was a brown-throated sloth with mange.

Chapter II
The United States of America

I covered American monsters in my 2019 book entitled, strangely enough, *Chasing American Monsters*. I won't cover old ground, so no repeats here. For those, you can check out that book where I detail more than 250 monsters from the US of A that creep, crawl, stomp, fly, and swim throughout the 50 states, including heavy hitters like Mothman, Momo, Champ, and the Hodag. Here, however, I'll give you a few I missed, at least one per state, like Bigfoot, a leprechaun, and a surprising number of water monsters.

Alabama

Alabama is probably most known for stereotypes, which is unfortunate. This state is beautiful, with Gulf Shore beaches, mountains and valleys, prairies, and an enormous cave system. Huntsville—a.k.a.,

Rocket City—was, and is, home to scientists who have developed the US space program since the 1950s. It was also the state where an African American bus passenger Rosa Parks refused to give up her seat to a white man, which was unheard of on December 1, 1955. Parks's bravery in the face of injustice was a catalyst for the American Civil Rights Movement. So, you're awesome, Alabama. Unless, of course, you're the Crichton Leprechaun. Then you're pretty friggin' dodgy.

The Crichton Leprechaun

The luck of the Emerald Isle graced the Mobile, Alabama, suburb of Crichton in 2006 when one citizen reported seeing a leprechaun in a tree.

A *leprechaun* in a *tree*.

The sighting made national news when Mobile's NBC affiliate WPMI drove to Crichton March 14, 2006, to investigate a claim local residents saw a diminutive, red-haired man in green clothing and a bowler hat up in a tree. The NBC news vehicle arrived on the scene to a crowd of leprechaun seekers.

The tree on Le Cren Street became a haven for onlookers days before the news channel showed up. In 2015, a news crew returned to Le Cren Street and found a man named Shun Thomas who claimed to be the person who first saw the leprechaun.

"Halfway through one beer, I'm looking through this tree here and I could see this image," Thomas told *The Mobile Real-Time News*. "He's still there."

Thomas said the leprechaun wasn't visible until he looked hard at the tree—then the Irish fairy appeared, slowly, in parts.

"It just looked like a leprechaun to me," Thomas said to *The Mobile Real-Time News*. "It was only an image from the chin to the top hat but that's what I thought about."

The WPMI report went viral and ended up on TV shows across the nation, including *Jimmy Kimmel Live*. A popular part of the broadcast was one resident shouting toward the crowd beneath the tree, "Who all seen the leprechaun, say, 'Yeah.'"

A sketch of the leprechaun appeared on the news segment, and later earned $1,100 from an eBay auction by the news station. The station donated the money to the American Cancer Society's Relay for Life.

So, a leprechaun? In Alabama?

We've all heard stranger.

Who all seen the leprechaun, say, "Yeah."

The Wolf Woman of Mobile

The April 8, 1971, headline read like a 1980s grocery store tabloid: "Is 'Wolf Woman' Sulking Around The City? Serious Area Persons Claim Seeing Creature." But it wasn't. The headline was from the *Mobile Press-Register*. The headline was serious.

Reports of encounters with an entity, with the head of a pretty woman and the body of a shaggy wolf, started to flow into the newspaper from the Mobile suburbs, unfortunately, on April 1. The beast was mainly reported from the Plateau neighborhood, and Davis Avenue, according to an article in *Mobile Bay Magazine*. By the time April 8 rolled around, the newspaper was prompted to publish a story on the creature, because fifty eyewitnesses called in claiming they'd encountered the Wolf Woman.

An obvious April Fools' Day hoax? Not so fast.

"One night, my father sent me to my grandmother's house to get something, and I had to walk by patches of woods," Antoinette Roberts Stewart told *Mobile Bay Magazine.* "Many of my friends lived all around us, but there were wooded lots between the houses. On that night, I swore I saw the Wolf Woman."

The Wolf Woman never attacked anyone, nor did she make much of a fuss, vanishing just as quickly as she'd appeared. However, she did put the fear of the dark into the local children.

There is one thing for sure. The drawing of the Wolf Woman published in that April 8 edition of the *Mobile Press-Register* isn't what I'd call friendly.

The Downey Booger

There's a booger in Winston County, Alabama. Well, there are probably lots of boogers, but I'm talking about a Booger, a Southern boogeyman that will eat your chickens if you're not careful.

It was during the 1800s, that part of the 1800s closing in on the new century, that two members of the local Downey family, cousins John and Joe, saw something as they rode their horses home at night through a thick forest of pine, according to a story by a descendant of the Downeys, Vera Whitehead, for the *Free State of Winston* website.

The Downey boys' horses snorted and reared up when they saw the Booger, an enormous, hairy, manlike beast. Then the horses bolted—in the wrong direction. The boys didn't get home until dawn. Of course, no one believed why they were late. Not until months later, when the whole family saw the Booger as it ran across the road.

As time went by, and more and more locals saw the Booger, it was too late for the Downey family to escape history. The monster was called the Downey Booger, and still is to this day.

There's an unincorporated village in the county named Booger Tree. The people in Winston County take their Boogers seriously.

Alaska

We'll get to Alaska, but I'm going to mention Texas first. Driving seventy miles per hour from the southernmost part of Texas to the northernmost part, the trip takes eleven hours. Alaska is more than twice the size of Texas. Alaska is 665,384 square miles (larger than 179 out of the 195 actual countries), most of those miles being picturesque wilderness, but the entire population is only 733,583, which is slightly smaller than the city of Seattle. One third of Alaska is above the Arctic Circle; it has the tallest mountain in North America (Denali, at 20,310 feet), bears (black, grizzly, and polar), and the most coastline of any state, which makes it a great place to find sea monsters.

Tizheruk

The Bering Sea lies between North American and Asia and was the site of the frozen land bridge anthropologists believe brought Asian peoples to the Americas thirteen thousand years ago (they were here a *lot* earlier, but that's a subject for another book). At its narrowest, the Bering Sea is fifty-one miles wide. That's all that separates the United States from Russia—fifty-one miles.

The Bering Sea is also where you'll find (or, won't find) Tizheruk (it's kind of slippery).

The Inuits of Nunivak and King Islands know of a stealthy sea serpent with a head as large as a grown man and a flipper at the tip of its tail. It can sneak to the water's edge and take people before anyone notices they're gone.

University of Chicago biochemist Roy P. Mackal addressed the Tizheruk in his 1980 book, *Searching for Hidden Animals*. He wrote that the Inuits would see the monster's long neck rise from the water by seven to eight feet and frighten them in their fishing boats.

Members of the US Coast Guard in Alaska confirmed the Inuit legends, because they'd seen the Tizheruk, too.

What no one could agree on was what the creature could be. A surviving dinosaur, a surviving prehistoric mammal, or a misplaced leopard seal (those live in Antarctic waters, not Arctic waters).

Whatever it is, the Tizheruk has been known to sink boats, eat humans who were standing around minding their own business, and generally make a nuisance of themselves. Shame on them.

Arizona

This Southwestern state is home to a natural formation I'm sure none of you have ever heard of—the Grand Canyon. It's a mile-deep slit in the earth. You know it? Yeah, I didn't think so. This state has mountains, deserts, and, in 1997, the city of Phoenix was visited by enough UFOs to create a flap that is still talked about. But nobody talks about the Lizard Man. A shame, he's a, you know, Lizard Man.

Lizard Man

On February 13, 2014, mountain bikers in the Old Pueblo Race saw a creature that should not exist. A Lizard Man.

A racer going by G. Johnson from Tucson told the *Tucson Weekly* he and two other racers were halfway through the race when they saw it.

"We had been riding for about...I don't know maybe nine hours, taking breaks every now and then," he told the *Tucson Weekly*. "When all of a sudden we see this long figure walking across the trail. He is maybe about six feet tall, very, very skinny, and it had an awkward gait, like a monkey...or a man with a disease, almost robotic, kind of."

And it was covered in scales. Johnson didn't know what to make of it.

"When you read these stories online or watch them on TV, well, you think, man, these people are crazy, on meds or something or in need of attention, but this has made me a believer," he told the newspaper. "There has to be more of them out there. If there's one, there's gotta be two at least right?"

He didn't go as far as saying the beast was the legendary Chupacabra, or even a space alien, but he was open to any logical explanation.

"All I'm saying is I have never seen anything like it in my life. But I am no biologist, so what do I know?"

Arkansas

This state is partly covered in mountains and forests, and partly covered in ghost stories (Eureka Springs, baby!). It's the boyhood home of President Bill Clinton, the only state with a public diamond mine (the biggest diamond in America was discovered there; it weighed more than forty carats), and, allegedly, the owner of Mexico Chiquito Restaurant in Little Rock gave America queso dip in the 1930s. Mmmm. Queso. The Water Panther, however, is something else.

Heber Springs Water Panther

A beast that's half mountain lion and half dragon has been a legend around Heber Springs for centuries. In the local Native American folklore (Ojibwe, Algonquin, Cree, Ottawa, Menominee, and Shawnee) it's called the Mishipeshu.

It lived in the deep parts of Heber Springs and is known to drag people to their death. Its size depends on the legends. Sometimes it's the size of a cougar, other times it's bigger than a bison. Regardless of its size, it always has antlers, a prehensile tail made of copper, and a "sharp saw-toothed back."

This fur-covered beast is known for its hellish scream most have only heard from deep in the forest. According to the 2007 book *Ozark Tales of Ghosts, Spirits, Hauntings, and Monsters* by W. C. Jameson, the locals said the creature's wail was "a cross between the cry of a panther and the scream of Satan himself."

The habitat of the Heber Springs Water Panther was flooded in the early 1960s, and sightings slowed down.

California

When it comes to state size, people are so flooded with images of Alaska and Texas, they forget about California. If California were its own country, it would be the fifty-ninth largest in the world. California sweeps roughly nine hundred miles along the West Coast from Mexico to Oregon. Known for LA and movies, it's easy to forget how diverse the geography is. With thick redwood forests, mountains, deserts, beaches, farms, California has it all. Famous people from the state are too numerous, and varied, to count (I'm looking at you, Richard Nixon and Snoop Dogg). However, it has its share of monsters, including the Fresno Nightcrawler.

HEBER SPRINGS
WATER PANTHER

Fresno Nightcrawler

In 2007, Fresno, a city of 550,000 residents between San Francisco and Los Angeles (and Las Vegas, if you want to make a triangle) became fascinated by home surveillance footage that depicted an unknown creature in a man's lawn. The creature—later, creatures—resurfaced in 2011, when video recorded two thin white entities with long legs and no arms taking a stroll in Yosemite Lakes Park, thirty-four miles away.

The video hit the internet, then, well, you know what happens. People went off the rails with speculation. Aliens, Chupacabras, demons, ghosts, a pair of pants... Watching the original video on YouTube, yes, it looks like a pair of white 1970s bellbottoms walking all by themselves.

Once the video went viral, a man in Montana reported seeing the creature, as well as a man in Poland. No, wait. The man in Montana didn't see the man in Poland. The man in Poland saw the Nightcrawler. English is hard.

The most detailed report came from a couple (who chose to remain anonymous) driving near Carmel, California (150 miles west of Fresno), who encountered the creature, about seven feet tall, near a place called Fort Hill.

"We were driving home," the wife told the *Highland County Press*. "After turning on Carmel Road, which leads to our road, we went around the curve by the Carmel church and then up a small incline and approximately ten feet over the incline and in front of our truck, the alien ran across the road and into the woods."

Her husband, who doesn't believe in aliens, told her what he saw.

"He wouldn't have admitted to seeing it if he hadn't been in shock," she said. "I had him draw it for me when we got to the house. He says it was asphalt gray, and about seven feet tall, no arms that he could see, but muscular in the legs area; no jawline, and its legs were bent backward and it leaned forward as it ran."

After 2020, when the man from Montana reported seeing the creature, its trail has gone cold. So, what was it? An alien? A huge bird? A hoax?

Who knows. If the Fresno Nightcrawler is really out there, it hasn't done harm to anyone, and probably just wants to find out where it left its shirt.

Elsie

Lake Elsinore, about sixty miles east of Laguna Beach, is the largest natural freshwater lake in Southern California, but that doesn't mean it's big. At three thousand acres with fourteen miles of shore length, it's dwarfed by a lake California shares with Nevada, Lake Tahoe, which has seventy-two miles of shoreline (and a lake monster, Tahoe Tessie. You can check it out in *Chasing American Monsters*). Lake Tahoe is 1,644 feet deep. Lake Elsinore has an average depth of twenty-seven feet, forty-two feet at its deepest.

But it also has Elsie.

Vague reports of a sea serpent in Lake Elsinore began in 1884. That changed in September 1934 when a Riverside County rancher, C. B. Greenstreet, claimed to have witnessed "the terrifying monster said to inhabit Lake Elsinore," according to an article in the *Los Angeles Times*. The *Times* of the day referred to Greenstreet as a "widely known and reputable valley rancher."

"It was a hundred feet long and had a thirty-foot tail," Greenstreet told the *Times*. "I know you think I'm crazy, but I saw it—we all saw it, my wife and two children—and my little girl was terrified."

The *Times* article also mentioned the lake monster has a long tradition in the legends of local Indigenous peoples.

Sightings of the creature continued in 1967, 1970, and 1992. However, Lake Elsinore dried up in 1930, and refilled by 1938. It also ran dry for years in the 1950s before refilling in the 1960s. And, you guessed it. There were no lake monsters dead in the dry lake bed.

The Lone Pine Mountain Devil

In the high desert on the east of the Sierra Nevada rises 12,949 feet of Lone Pine Mountain. The mountain, and the town of Lone Pine, is named after a pine tree that once stood at the entrance to (of course) Lone Pine Canyon (a flood felled the tree in the 1800s). More than four hundred movies, one hundred episodes of television, and an endless number of commercials have been filmed in, and around, Lone Pine Mountain. And, for Macintosh users, the mountain is the desktop photograph of the Mac operating system Sierra.

Despite the beauty and the blessing of the Apple corporation, the mountain is also home to the Lone Pine Mountain Devil.

This gigantic creature, part bird, part lizard, has four wings and deadly sharp claws. When it attacks, it goes for the face of its prey, working its way down to its trunk; the Lone Pine Mountain Devil doesn't just kill for food, it kills for fun.

The most discussed story of this Devil takes us into the late 1870s when a wagon train of Spanish settlers was attacked by a monster in the Sierra Nevada. Padre Justus Martinez was the sole survivor.

One night on the journey, the settlers camped in the mountains and stayed up late in the night to celebrate the fact their journey was nearly complete, when winged beasts "damned by the good Lord" fell upon the party, slaughtering them all, Martinez wrote in his journal. Martinez had pitched his tent on the outskirts of the camp, away from the revelry, and was spared, according to multiple sources. "May the forgiving Lord not abandon their souls, which were taken from them into the depths of hell!"

On November 12, 2013, *Sierra Wave Media* posted a letter to the editor about the Lone Pine Mountain Devil. Here is part of the letter:

"I wish to seek comment on a reemerging phenomenon that few valley people I speak with are aware of; the Lone Pine Mountain Devil. On a recent visit to Independence (a small town fifteen and a half miles north of Lone Pine), I noticed a Native basket decorated with a batlike winged creature. I had been exposed to Lone Pine Devil stories through reading folklore yet never through speaking with a local. The basket struck me as a sort of historic document and rekindled my interest.

"I wonder if any of your readers know of the stories or have seen this Devil themselves."

If you're going to ask for input, you should know what kind you'll get.

Comments:

- It's Obama's fault.
- I live in Lone Pine, and when I mentioned this to people around here today, no one has heard of it…but it is an interesting story and folklore.
- This is about as real as the tooth fairy. We have actually be(en) pretty gentle with the ribbing.

The author of the letter suggested the Lone Pine Mountain Devil sounds similar to *Sinornithosaurus milleni*, a Cretaceous feathered dinosaur paleontologists have said was carnivorous, and probably hunted in packs.

Even though *Sinornithosaurus milleni* went extinct 125 million years ago, it seems like a good fit.

Colorado

Oh, Colorado. The Centennial State. The state with the most 14ers (mountains taller than 14,000 feet. It has 58). The birthplace of Duane "Dog the Bounty Hunter" Chapman. The first state to legalize recreational marijuana. It also apparently has a Bigfoot infestation. Like mice, if you see one Bigfoot, there are probably five or six in your basement, attic, or hiding in the walls of your house.

The Durango Bigfoot

A couple from Wyoming were hoping to spot elk on an October 2023 excursion aboard southwestern Colorado's Durango & Silverton Narrow Gauge Railroad, when they saw something they didn't expect—Bigfoot.

At least they think it was Bigfoot.

The Durango & Silverton Narrow Gauge Railroad provides a scenic trip that runs for forty-five miles between the small historic towns of Silverton and Durango (Durango is about twenty-seven miles north of the New Mexico border). The train takes passengers through the San Juan Mountains and stretches of nature inaccessible to cars.

The couple, Shannon and Stetson Parker, didn't know what to make of the creature at first.

"My husband sees something moving and then can't really explain it. So he's like, 'Bigfoot,'" Shannon said to the *New York Post.* "It was at least six, seven feet, or taller. It matched the sage in the mountains so much that he's like camouflaged when crouching down."

A nearby passenger identified only as Brandon, recorded the creature with his mobile phone. The video went viral.

"Brandon, the guy sitting next to Stetson on the train grabs his phone and starts recording," Shannon posted to Facebook. "Y'all, out of the hundreds of people on the train, three or four of us actually saw, as Stetson says in the video, the ever elusive creature Bigfoot! I don't know about y'all but we believe!"

The curator of Felton, California's Bigfoot Discovery Museum, Michael Rugg, told the *New York Post* the Bigfoot "had a strong possibility of being a guy in a suit. It didn't look right to me."

To Rugg, the creature didn't appear as muscular as Sasquatch is usually reported, however, "There's not enough detail to be able to judge."

According to the *Durango Herald*, not even the Parkers are claiming it's 100 percent Bigfoot.

"I don't know. I mean, I just think we saw what we saw," Shannon Parker told the *Herald*. "I don't know that I've seen anything that's definitive."

The *Herald* reported that, in the past, people have worn Bigfoot costumes to entertain people aboard the sightseeing trains between Silverton and Durango. For now, no one has stepped up to take responsibility.

So, the video could show a hoaxer, a hunter in a ghillie suit, or an actual Bigfoot. Ook.

Blue Dilly

Lake Dillon—otherwise known as the Dillon Reservoir—has a surface elevation of 9,017 feet and a shore length of nearly twenty-seven miles; it's west of Denver near the towns of Dillon, Frisco, and Silverthorne. The reservoir area is popular for boat racing, snow skiing (the resorts of Keystone, Breckenridge, and Copper Mountain are all nearby), snowmobiling, hiking, camping, and fishing. The lake is well stocked with salmon, trout, char, and smallmouth bass. People "can't swim, scuba dive, or water-ski on the lake due to the cold temperatures and water quality concerns," according to denverwater.org.

However, that isn't a concern for the lake's water monster, Blue Dilly.

Wait. Let's take a step back. Maybe, a few steps back.

During the Western gold and silver rush, a hopeful named Ruben Spalding found gold at the headwaters of the Blue River (chances are the Indigenous peoples already knew it was there). Over the next twenty years, the white population in Summit County exploded from 100 to around 5,500. By 1900, the local government knew the area's residents needed better access to water, and the plan to build a dam was floated (see what I did there?). It took until 1963, but an earth-filled, 5,888-foot-long dam was completed, and water from the Blue River Basin filled an area that used to hold the town of Dillon.

So, if the lake is man-made, could it seriously hold a lake monster?

The Dillon Reservoir sinks to about 320 feet at its deepest, a suitable size for a lake monster, but Blue Dilly isn't a normal lake monster. Since 1965, people have reported seeing a freshwater

BLUE DILLY

manta ray—an enormous freshwater manta ray—that tops out at sixty-foot long, with a thirty-foot tail. Damn.

Manta rays, also known as devilfish, can live up to fifty years, and are intelligent, having the largest brain-to-body ratio of any fish on the planet—but they're ocean fish. Mantas have been seen in fresh water in places such as Hawaii if they're hungry, or are evading predators. The only freshwater rays are the Amazon River stingrays, although they're just a tad bit smaller than Blue Dilly's sixty-foot-long body by being, at most, eighteen inches long with a twelve-inch tail. They only live from five to ten years, so unlike the solitary Blue Dilly, they'd need a large breeding population to survive.

And heat. They'd need heat. The frigid Lake Dillon has a high of 52°F and a low of 22°F, whereas the average water temperature of the Amazon River is in the 80s.

Hoax, or unknown species? Who knows. If you think you've seen something large and threatening in the Dillon Reservoir, report your sighting at Bluedillonmonster.com.

However, if you saw something large and threatening on the lake in April 2020, an artist constructed a Loch Ness–like monster atop the frozen waters about three hundred feet from the Dillon Marina. Don't report that. I mean, unless you really, really want to.

Teihiihan

Fast, strong, and hungry. The diminutive Teihiihan are similar to legends of angry little people from across the world. This race of dwarves in Colorado and Wyoming terrorized the Arapaho peoples. They are said to have dark skin, superhuman strength and speed, can become invisible (I'm looking at you, Pukwudgies and Menehune). They're also cannibals (Ebu Gogo, anyone?).

Apart from their physical prowess (and that whole invisibility thing), Teihiihan were notoriously hard to kill because they possessed the ability to remove their hearts and hide them. With no heart to pierce, the Teihiihan couldn't be defeated in battle.

One Arapaho legend tells the story of a young man kidnapped by a Teihiihan who takes him home for its family to eat. However, the man outsmarts his captors and asks the family about the hearts scattered around their home. When they tell him the significance of those hearts, he grabs a knife and stabs each heart, killing the entire family.

Another story has area tribes banding together and wiping out these cannibal dwarves. Either way, the Teihiihan aren't around to bother us anymore.

Connecticut

One of the original thirteen colonies, this small New England state has a proud history, one that involves education, and alcohol. Yale University is in Connecticut, the first publicly funded US library was opened in the state, and Connecticut native Noah Webster wrote the first American dictionary there (he's the reason we don't spell *color* with a *u*). Connecticut, and nearby Rhode Island, were the only two states not to ratify the 18th Amendment legalizing Prohibition. Maybe the booze was to deal with all the vampires. Bottoms up.

Jewett City Vampires

Tuberculosis is an awful disease.

From the 1600s to the 1800s, tuberculosis (TB) "caused 25 percent of all deaths" across Europe and the United States, according to the Centers for Disease Control and Prevention (CDC).

Until COVID-19, TB was still the "deadliest infectious disease in the world," as per the CDC.

"No spitting" laws were passed in major cities to help control spread of the disease, because TB is passed through saliva, according to National Public Radio.

But that didn't help the Ray family of Jewett City, Connecticut, because in May 1854, the disease hit hard. However, in New England in the 1800s, healthy people growing pale and wasting away wasn't caused by a disease. It was caused by vampires.

According to the *Register Citizen* in New Haven, after doctors couldn't help the family, they dug up the bodies of their two sons and set fire to them, believing those boys were vampires who rose at night to feed on other members of the family.

"People were frightened. It was a final effort to save the living," state archaeologist Nicholas Bellantoni told the *Register Citizen* in 2008.

The Rays weren't alone in this belief, of course. In the 1990s, twenty-nine unmarked graves of the Walton family were discovered near Hopeville, Connecticut. One man, buried in the 1790s, had been exhumed about five years after his death, and his head had been removed and placed on his chest, looking down the length of his body. His femurs were placed in an X below the skull—the skull and crossbones, according to the *Register Citizen*.

That skeleton and two more (a woman and a child) all bore the marks of having TB.

Where was the site? About two miles from the Ray family farm.

Delaware

The second-smallest state in the union, Delaware packs a punch—or a peck. Its state bird, the blue hen chicken, was chosen because that breed was used for cockfighting, which, in the early days of the state, was a popular sport. Delaware still has chickens. Lots of them; at two million birds, there are twice as many chickens as residents. It also has the Prime Hook Swamp Creature, which is not a chicken.

Prime Hook Swamp Creature

The Prime Hook National Wildlife Refuge is ten thousand acres of wildlife habitat. About 80 percent is wetlands (salt and freshwater) that includes angry-sounding landmarks, such as Slaughter Beach and Broadkill River. Yikes.

Then there's the creature.

According to the article "Discovering Delaware's Daunting Cryptids" by Wade Beaumont, a woman named Helen J. saw a creature about three feet tall with "long, spindly legs and a tan body." Its face was like a pug's, with small ears and angry eyes. She estimated this beast weighed about thirty pounds. Helen's daughter also saw the creature once, as did a nearby store owner.

What was it? A deformed dog, or coyote? A space alien's pet Daggit? Or a cryptid?

Florida

Florida! Woot!

Okay, let's get Florida Man out of the way. Yes, Florida Man assaulted a person with spaghetti. Yes, Florida Man tried to order a blunt at a fast-food restaurant. Yes, police arrested Florida Man for returning used enemas to a pharmacy. And, yes, Florida Man left his ten-month-old baby in the care of a babysitter who was a dog. Why is Florida Man so crazy? He's not. Although Florida is an, um, interesting state, the reason Florida Man is a meme has to do with Florida having the most lax public records laws in the country. It also has Tarpie.

Tarpie

Lake Tarpon is a big lake with a surface area of 2,534 acres, and it's absolutely filthy. Although it's a "fishing lake" (known for large-mouth bass), Tarpon is on the EPA's "stay the hell out of the water" list because of pollution. You can swim there...if you want, but you probably shouldn't want to. Honestly, I wouldn't want to eat the fish, either.

Then, there's Tarpie.

The Lake Tarpon monster, Tarpie, is reported as a fifteen- to thirty-foot-long reptile living in the lake that may be part alligator, part manatee, part fish, or a dinosaur, or part—whatever.

This beast not only swims in the lake, it walks on its banks—on two legs. Some claim it's been around for before Europeans came to American shores, but there are, of course, no records. However, legends claim it has huge claws, a head like an alligator, and fur over reptilian skin.

An internet post from a man named George claims he saw the creature in 2006. After piloting a boat into Brooker Creek (the only creek that enters Lake Tarpon), the man not only saw an alligator about twelve feet long, they saw a water creature he said looked like a snake a foot thick and about thirty feet wide.

Doesn't sound like a bipedal, semiaquatic, dinosaur-like lake monster to me, but given the invasive snake species in Florida,

TARPIE

such as the Burmese python that can grow as long as twenty-two feet, he may have seen something real.

Tarpie's not the only monster in Lake Tarpon, in 2013, officials captured an eleven-foot-long crocodile in the lake. It wasn't supposed to be there.

Georgia

Georgia, the largest state east of the Mississippi River, has Atlantic beaches, mountains, and farms. It's home to the Masters Golf Tournament, the first college in the world that granted degrees to women (Wesleyan College), is known for peaches (but produces more pecans), and, for some reason, once made it illegal to keep a donkey in a bathtub. It may also have a velociraptor.

Georgia Velociraptor

An eighteen-year-old and his grandfather were hunting deer the early morning of July 25, 2008, when they encountered what they thought was a velociraptor, a birdlike dinosaur that went extinct in the late Cretaceous period.

The young man, Y. Phillips, wrote of the encounter on Thoughtco.com.

"We heard an unusual noise we never heard before on our many hunting trips. Grandpa looked at me and listened. Then he raised his finger in front of his mouth to show me that we shouldn't make more movements. I heard a lot of movement and more of the noise. I can't really describe the sounds, but I sure can describe what I saw, even when it was pretty dark. We just kept listening to the sounds as suddenly something came walking slowly out of the bushes and onto the road maybe 150 yards in front of us. My eyes got really big, and at that moment I wasn't even scared, just amazed to see this creature."

Phillips wrote the creature looked like a velociraptor from the popular *Jurassic Park* movies.

The monster had "a long, stiff tail, walked on two feet and had short arms. It looked lizard-like and had a huge claw on both of his feet and smaller claws on his arms."

Phillips and his grandfather remained still so as not to alert the creature. It appeared to sniff the air before running into the bushes. They waited, and waited, and the beast never returned, so they got back into their pickup and left.

"Since that encounter," Phillips wrote, "I believe in creatures that science doesn't know about. That's my story, as odd as it sounds. I know what I saw."

Hawaii

Hawaii is a long way from everywhere—twenty-four hundred miles. Being isolated in the middle of the Pacific, Hawaii's biosphere is unique. That is until Europeans arrived on the islands on January 20, 1778, and messed everything up. Now the islands have cane toads, the brown tree snake, Axis deer, and enough feral chickens to open a restaurant chain. There's a seventy-mile-wide volcano on the islands, it's the only state in the tropics, and on July 2015, all Hawaiian islands banned plastic bags in their stores. Good for them. It also has a pig god.

Kamapua'a—The Pig God

When a young couple on the island of Oahu had a baby, they didn't expect parenthood to be such a challenge. As the boy Kamapua'a grew, he became increasingly troublesome, getting into mischief after mischief. Then, when Kamapua'a discovered he had the power to change from a boy into a pig, he flaunted the talent to the villagers, and they began to worship him as a god.

This pleased Kamapua'a, and, when he became a man, it inflated his ego even more. He grew to be strong, and agile, and with this he also grew to be loud, and hungry, and took his aggression out on the humans around him.

But Kamapua'a was also handsome, and could cause the crops to grow, so all the young women of the surrounding villages fell in love with him. However, Kamapua'a didn't want them. He fell in love with a woman named Pele (the volcano goddess, not the famous Brazilian soccer player). Pele, however, could see through the charisma of the Pig God, and shunned him, at first. However, Kamapua'a's charm finally got to Pele, and they began a romance.

It didn't last because Kamapua'a was (if I can get away with saying this) pigheaded.

During a fight, Kamapua'a tried to put out Pele's fire, and Pele banished the Pig God to the other side of the Big Island. Today, people can tell which part of the island is which. The lava fields are, of course, Pele's, and any land with crops are Kamapua'a's.

Idaho

Part of the Pacific Northwest with Washington, Oregon, and Canada's province British Columbia, Idaho seems a bit out of place with these forested lands. The Rocky Mountains run through Idaho, and this state is home to the deepest gorge in the US at 7,993 feet, Hells Canyon, and shares the Owyhee Desert with Nevada; however, it's far from odd man out. National forests cover about 40 percent of the state. Idaho is one of only two areas in the world home to seventy-two different types of gemstones, and it's the site of daredevil stuntman Evel Knievel's attempt to jump a rocket cycle over the Snake River Canyon (he failed, gloriously). Idaho is the highest-potato-producing state. It's also known for huckleberries, whatever those are. So, could it also be home to a lake monster?

Paddler

Pend Oreille Lake, about 85 miles east of Spokane, Washington, is the fifth-deepest lake in the United States, the deepest spot nearing 1,150 feet, which is around 200 feet deeper than Lake Michigan. It is 43 miles long and has a surface area of 148 square miles.

That's a lot of water.

The lake was formed during the ice age as glaciers pushed down from Canada, and made larger by the US Army Corps of Engineers when it constructed a dam at Albeni Falls near the town of Newport. After the Japanese attack on Pearl Harbor in 1941, the Navy built the second-largest naval training facility in the world at Pend Oreille Lake and tested submarines there. Because of the size and depth of the lake, the sound density was similar to the open ocean.

Some locals say it's home to a fella named Paddler.

Although there were early reports of something strange in the lake in 1944 (prime submarine testing time), the first substantial sighting of a lake monster happened in 1977. A girl swimming near the lake claimed she was attacked by a monster the local newspapers dubbed the Pend Oreille Paddler. The beast was said to be about twenty feet long, which is on the large size of the biggest freshwater fish in the United States, the white sturgeon. However, to put that into perspective, the largest white sturgeon ever caught in Idaho was in 2022 at the C. J. Strike Reservoir (five hundred miles south of Pend Oreille Lake); it was ten feet, four inches long, a far cry from Paddler's size.

A professor from nearby North Idaho College, Jim McLeod, spent years chasing the legend of Paddler, launching the investigation Crypto Quest 1984, which garnered a slew of witness testimony, but no lake monster. McLeod said those sightings could be attributed to an abnormally large sturgeon, or submarines.

However, the year after Crypto Quest on the lake, a teacher from nearby Coeur d'Alene, Julie Green, was spending Memorial Day at the lake with friends when they saw something strange near their boat.

"There was clearly something in the water ahead of us that was undulating, coming in and out of the water," she told the local press.

The object was "gunmetal gray" and as large as the twenty-two-foot boat the group rode. They tried to chase the thing, but it was too fast.

In 2007, a photographer for a local newspaper, the *River Journal*, published a photograph of humps on the water that could be Paddler.

Well, like I said, it's a big lake. Anything's possible.

Residents of the Lockridge area flew into a panic. However, it was short lived. Sightings of the monster tapered off, and life soon went back to normal.

Illinois

Illinois, contrary to what Hollywood tells us, is more than just Chicago. This state is filled with farms, plains, and forests. The first McDonald's Restaurant opened in Des Plaines, and Twinkies were invented in River Forest. Illinois produces more nuclear energy and pumpkins than any other state. Those may, or may not, have something to do with the Lake Michigan Sea Serpent.

LAKE MICHIGAN
SEA SERPENT

Lake Michigan Sea Serpent

Back in 1893, officers at the Army post Fort Sheridan (now a residential area) reported seeing a sea serpent in Lake Michigan and were so terrified they vowed to give up liquor, according to the *Chicago Tribune*.

However, the sightings continued.

In 1899, a woman living in Lincoln Park told the *Tribune*, "I looked through my opera glasses and I could see it was not like a boat."

The first sighting was in 1817 and continued until the 1930s. The monster has been described as snakelike, or eellike, swimming just offshore. Its head looks like an alligator's.

It's not seen often, but at least often enough to grab an identity like the Loch Ness Monster, or Champ of Lake Champlain. Some people call it South Bay Bessie, but, then again, other people might not know what they're talking about.

Indiana

This Midwest state knows quite a bit about corn. Half the crops produced in the state are corn, and Indiana produces 20 percent of the nation's supply. It has a town named Santa Claus that receives thousands of letters to Santa each Christmas—all get a reply. The state colors are blue and gold...wait. Does that mean Indiana is in Ravenclaw? The first successful goldfish farm opened there in 1899. Cool? Nah. Meshekenabek would eat them.

Meshekenabek

With a surface area of 40 square miles and a maximum depth of 162 feet, Lake Manitou is big enough for a monster. With a twelve-letter name like Meshekenabek, it had better be.

The local Native American Potawatomi tribe shunned the Devil's Lake, believing it to be possessed by an evil spirit. They didn't fish there, bathe there, or probably make s'mores there.

The evil spirit, embodied in the monster Meshekenabek, was said to be a sixty-foot-long serpent with the head of a cow. The Potawatomi warned white settlers about the monster, but, of course, they didn't listen.

Because of the monster, surveyor Austin W. Morris couldn't keep men employed to plant surveyor flags near the water, and Fulton County's first blacksmith, John Lindsay, personally saw the beast.

"The head being about three feet across the frontal bone and having something of the contour of a beef's head, but the neck tapering and having the character of the serpent. Color, dingy with large yellow spots," the *Fulton County Post* reported.

The monster apparently raised its head out of the water about two hundred feet from shore as Lindsay watched.

According to a July 1838 edition of the *Logansport Telegraph*, fishermen saw a sixty-foot-long snakelike monster in the lake.

As the legend of Meshekenabek spread, ocean fishermen came to Indiana to hunt the beast. They all failed.

Several large fish, such as an enormous buffalo carp, and a 116-pound spoonbill, caused people to claim the monster had finally been hooked, but true believers know those small fish weren't the Devil's Lake Monster.

Iowa

Iowa sits on top of Missouri like a hat. This state is mostly plains and farms and has more hogs than people. According to the US Department of Agriculture, for each Iowa resident, there are 7.3 pigs, which makes it the logical state for the annual Blue Ribbon

Bacon Festival. It's the only state whose borders are two navigable rivers (the Missouri and Mississippi). Natives include painter Grant Wood and actor John Wayne, and the East Okoboji Lake Monster.

East Okoboji Lake Monster

As part of the Iowa Great Lakes, East Okoboji Lake is, well, great. At 7.43 square miles, it's a haven for fishermen and boaters. With an average depth of twenty-two feet, it may not be deep enough for a monster, but you can't blame a monster for trying.

In the early 1900s, a couple at the lake told the *Vindicator and Republican* from nearby Estherville, "They saw something, or rather saw where something was. They have no idea what it was. It might have been a sea serpent, or it might have been some kind of a fish that had grown to unusual and extraordinary size."

The couple was boating when the nearby water began to stir.

A living object swam toward them, large enough to overturn their boat. The waves it made almost did just that.

"He does not pretend to know what it was and declares he would not have believed there was such a creature in the lake had he not seen with his own eyes the commotion made by it," the *Vindicator and Republican* reported.

But, as for reports, that's it. If something exists in the lake, it's been there a while—the lake was formed by a glacier that tore through Iowa twelve thousand years ago.

Kansas

Nestled between Missouri, Nebraska, Colorado, and Oklahoma—all states with mountains (yes, even Nebraska)—the state of Kansas has the distinction of being flat. Really flat. Pancake flat. This is not a joke. This is not hyperbole. In 2003, geographers at Texas State University and Arizona State University, obviously bored with life, measured how flat Kansas actually is, and compared that to the flatness of a pancake. If 1.000 means perfectly flat, a pancake rated 0.957; Kansas rated 0.9997, which the scientists said was "damn flat." Kansas is also known for wheat production, having the USA's largest amount of acreage designated for planting wheat. And, of course, it has Sinkhole Sam.

Sinkhole Sam

Surprisingly enough, Inman, Kansas, a town of 1,320 residents in the dry and dusty Great Plains, was once home to a lake monster. Back in the ancient past (you know, the 1920s) a cluster of freshwater lakes dotted the fields around Inman, the largest being Lake Inman, a 160-acre lake, which was the largest natural lake in the state. According to the *Hays Post* of Hays, Kansas, these lakes attracted duck hunters from all over.

The lakes were eventually drained, presumably for more farmland, but the deepest depressions left, which still held water, were called sinkholes. That's where Sam comes in. People began reporting a sea serpent in the biggest of these sinkholes named the Big Sinkhole (A-plus for creativity).

The first sighting of Sam was by fishermen at the Big Sinkhole. The second was from two men from Inman, Albert Neufeld and George Regehr. According to the *Post*, Neufeld fetched his rifle and aimed a few errant shots at the giant water snake. They said the snake was fifteen feet long, and thick.

Since stories like this spread from the local newspapers to national via the *Associated Press*, the countryside was filled with cars from all over because people wanted to get a peek at Sinkhole Sam.

What was Sam? Locals thought draining the lakes filled an underground cave Sam had been sleeping in. Sam woke, then came out to play. Scientists labeled the creature a Foopengerkle because they didn't think Sinkhole Sam existed, so he deserved a silly name.

A headline in the November 23, 1952, edition of the *Salina (Kansas) Journal* read, "Science Set Back Years: Monster Turns Out to Be a Plain Old Foopengerkle."

Sorry, Inman. The press can be harsh.

Kentucky

The Commonwealth of Kentucky is one of four states labeled as such, along with Massachusetts, Pennsylvania, and Virginia. This southern state is the home of bourbon whiskey and is absolutely beautiful. With places like Red River Gorge, Cumberland Gap, and Mammoth Cave (the longest cave system in the world at 365 miles), Kentucky is a haven for nature lovers. Kentucky's also known for monsters, which I wrote about in *Chasing American Monsters*, and with half the state's land dedicated to commercial forest land, there's plenty of room for creatures like the Spottsville Monster to hide.

Spottsville Monster

In 1975, a former coal town of 325 people, about twenty-two miles west of Owensboro (the fourth-largest city in Kentucky), was visited by a big, hairy, humanlike beast.

That year, the Nunnelly family, who lived just outside town on a farm, noticed their animals were disappearing. A neighbor, hunting in the nearby woods, encountered an eight-foot-tall, hairy bipedal creature, and, aware of the Nunnelly's missing animals, warned them the monster may be the cause.

More on them later.

The neighbor encountered the beast again, and again, and discovered the hairy fella didn't behave like a terrestrial animal. The creature, that would soon become known as the Spottsville Monster, didn't leave footprints, could vanish into thin air, and encounters with it were often accompanied by UFO sightings.

Wait. What?

The Nunnellys (I told you I'd come back to them) saw the creature for themselves on October 20, 1975. According to the *Gleaner* (the newspaper from nearby Henderson, Kentucky), the Nunnelly children finally saw the hairy, tall, dark green, humanlike monster outside their home.

Then the monster wouldn't leave them alone.

More farm animal deaths, howls, the hulking beast lurking in the shadows—the Spottsville Monster terrorized the Nunnelly family. The Destination America program, *Monsters and Mysteries in America*, featured this story, and Lonnie Hawks, a local Kentucky Bigfoot researcher, said, "This family lived this. They finally got sick and tired of this thing aggravating them and they moved to town."

There are still a lot of questions about the Spottsville Monster. Interviews reflect that the Nunnellys believe they encountered a UFO-buddy-vanishing hairy beast.

Louisiana

A blend of cultures—American, African, and French—makes Louisiana unique, and not just because of its Cajun and Creole cuisines. Because of its French heritage, Louisiana calls its counties parishes, and, instead of basing its civil laws on British common law (as has most of the nation), Louisiana civil law is based on the Napoleonic

Code. Known for jazz, Mardi Gras, and the Turducken (invented by New Orleans chef Paul Prudhomme), Louisiana is also home of the Wild Girl of Catahoula.

The Wild Girl of Catahoula

In the late 1800s, the people of Catahoula Parish began seeing a naked girl running through the woods.

The first report came in 1886, when Jack Francis's teenage daughter (girls apparently didn't have first names back then) saw a "Wild Girl, perfectly nude, with long black hair," according to the *Jena Times-Olla-Tullos Signal.* When the naked girl saw Miss Francis, she ran off.

More sightings began to come from the parish. People on horseback occasionally approached the girl and gave chase, but the Wild Girl, despite a deformed foot, was fast—really fast—and always managed to escape.

The identity of the Wild Girl was pinned on the daughter of a poor woman named Duck who used to "tramp through the country with three children," one of whom had a clubfoot. Locals speculated Duck abandoned the child in the woods.

In 1887, a report in the *Times-Picayne* of New Orleans said that the Swilley family witnessed the girl kill a goose and run off with it.

Hunting parties spread out through Catahoula Parish to look for the girl, but the only evidence of her they could ever find were bare footprints, one of which showed a clubfoot.

In a later encounter that year near Hemp's Creek, witnesses said the girl "was one of the most ferocious-looking beings that the human eye had ever cast upon." When they approached her,

WILD GIRL OF CATAHOULA

she dashed away, leaping like a deer. They described the girl as sixteen years old, about four feet, six inches, 125 pounds, and walking with a limp. She "was clothed with nothing but what nature gave her."

In September, Captain J. M. Ball, John C. Goulden, M. W. Calvitt, and Charles Goldenburg were fishing in Grant Parish—fifty miles away. They saw her carrying a knife and a pig and told the *Times-Picayne* she was "covered with hair, varying in length on different parts of her."

By 1890, reports changed. Members of the Hardtner family were taking a twenty-one-mile buggy trip from Fishville to Pineville when they saw a girl "wearing a faded homespun dress and was barefooted." She ran from the buggy "at a speed, all say, they never saw a human being run at."

The final sighting of the Wild Girl of Catahoula occurred in July 1891. Witnesses said she was "very tall and powerful, covered with hair, and carrying a knife or a sword."

She disappeared after that. Whether she was real, or not, is lost to history.

Maine

Maine, a place of trees, moose, lighthouses, lobsters, and, if the most famous Maine resident author Stephen King is right about this, more evil than the rest of the states combined (probably not. It's too pretty). The state produces blueberries, lots and lots of blueberries, exporting more than 90 percent of America's total. There are 4,000 islands off the coast of Maine, and, including those islands, Maine has more coastline than California (3,478 miles to 3,427). With all that Atlantic coast, there has to be a sea serpent there, right?

Casco Bay Sea Serpent

An early sea serpent sighting in Maine occurred in 1779, when an ensign aboard the ship *Protector* saw a beast in Penobscot Bay, according to *Emergence Magazine*. The ensign, Edward Preble, boarded a longboat, rowed toward the long-necked creature, and aimed a rifle at it. His shot missed, and the creature dove back into the depths.

Unfortunately, or fortunately, however you want to view it, many of these early sightings are the same. *Emergence* reported a similar sighting in Portland Harbor in 1818, and in 1836, a Captain Black commanding the ship *Fox* saw a long-necked, snakelike sea serpent off the coast of the island Mount Desert Rock. An identical serpent was seen off Wood Island Light in 1905, and sailors aboard the steamer *Bonita* saw a creature at least eighty feet long next to the ship in 1910.

Another thing these sightings have in common is that people have attributed them all to the Casco Bay sea serpent.

Casco Bay is in southern Maine and hosts its biggest city, Portland, along with Maine's most famous lighthouse, Portland Head Light.

Although sightings haven't been discussed since the early 1900s, Loren Coleman, cryptozoologist, author, executive director of the International Cryptozoology Museum in Portland, and the person who wrote the foreword to this book, interviewed a fisherman named Ole Mikkelson who encountered the Casco Bay sea serpent.

Mikkelson and his first mate were fishing off Cape Elizabeth in June 1958 when they saw "a large serpent creature colored like a flounder" coming close to their boat, according to *News Center Maine*. Before it dipped beneath the water, it would turn its head toward any foghorn that would blat across the water.

Maryland

What Maryland lacks in size (it's the ninth-smallest state), it more than makes up for in awesomeness. Maryland is known as the Free State because when the state constitution took effect on November 1, 1864, it abolished slavery within its borders. The *Maryland Gazette* began publishing in 1727 and is *still* publishing. The first post office system opened in Baltimore in 1774, the first municipal water company was started in Baltimore in 1792, St. Francis Academy opened as the world's first dental school in 1828, and the world's first telegraph line connected Baltimore with Washington, DC, in 1844. It also has werewolves.

Dwayyo

The hubbub began with an article in the *Frederick News-Post* on November 27, 1965.

A local man named John Becker (police could find no one in the area by that name) went outside his house on Fern Rock Road (police could find no road in the area by that name) out by Gambrill State Park (a real place) to locate the source of some odd noises when he saw something unbelievable. It was a creature "big as a bear, had long black hair, a bushy tail, and growled like a wolf or a dog in anger," he told the *News-Post.*

Then Becker did something opposite of a survival technique by walking closer. The beast stood on two legs and attacked the man, who fought back. The creature eventually fled into the nearby trees, and Becker went back inside and called the police (although doing that first would have probably been a better idea).

As sightings such as this go, right after the newspaper published the article, more and more people reported seeing the creature. But, with the increased publicity came an identification of the monster—the Dwayyo.

The Dwayyo is a six-foot-tall, bipedal, wolflike beast who lives in the Maryland woods. It was first reported in 1944 in Frederick County, Maryland. Residents saw enormous canine footprints and heard terrifying screams from the woods.

Back to the Becker account. For months after, residents of a neighborhood that backed up to a forest reported hearing a baby cry from the wood line that would eventually turn into a woman's shriek. One woman claimed to see a huge, doglike creature chasing cows.

During this initial 1965 fervor, dozens of students from the nearby Frederick Community College planned a Dwayyo hunt, but it didn't materialize because no one was brave enough to go into the woods at night in search of a big black wolf that walks on its hind legs. College students are smart.

So, the 1965 witness used a false name, the locals who reported seeing the monster didn't have physical proof, and police, such as Sgt. Clyde B. Tucker who took the initial report, said there's no such thing as a Dwayyo, but wrote up a report anyway because, "Sure enough, some hunter would call the police barrack saying he had just killed one."

Sure enough, a hunter who wanted to remain anonymous told the newspaper he'd seen the monster.

"My dogs started chasing something, and I saw it was black, but I didn't think too much of it, believing it was a dog or maybe even a bear," he told the *News-Post*. "However, after reading the newspaper article, I'm not too sure it wasn't a Dwayyo. It trotted like a horse. I don't know what it was, but I'm going to look for it this week."

The sightings continued.

In 1966, a man identifying himself as only Jim A. was camping at Gambrill State Park, about six miles outside Frederick, when he saw a "shaggy two-legged animal the size of a deer with a triangle-shaped head and pointed ears and chin." Like earlier reports, Jim A. said the creature screamed like a terrified woman.

In 1976, two men driving near Thurmont, a small town sixteen miles north of Frederick, claimed that a monster ran in front of their car. The beast was "at least six feet tall, but inclined forward since it was moving quickly. Its head was fairly large and similar to the profile of a wolf." Its rear legs were thick, and heavily muscled, like a kangaroo's.

In 1978, park rangers saw a "large, hairy creature running on two legs" near Cunningham Falls State Park.

A werewolf? Maybe.

Massachusetts

Massachusetts has always been a home to politics, from its pivotal role in the Revolutionary War, to being the birthplace of four presidents (John Adams, John Quincy Adams, John F. Kennedy, and George H. W. Bush), it has had a say in how our country is run. It is the sixth-smallest state, but has the highest population of all New England states with seven million people. It's most known for the origin of Thanksgiving, and the hanging of nineteen witches in the seventeenth century. The sports of basketball and volleyball were invented there. It's also the location of the Bridgewater Triangle.

Thunderbird of the Bridgewater Triangle

The Bridgewater Triangle is a spot of mystery, including hauntings, Bigfoot, UFOs, enormous lizards, and other cryptids. This triangle (think Bermuda Triangle, Alaska Triangle, the Devil's Triangle in the Sea of Japan, etc.) is between the towns of Freetown, Rehoboth, and Abington. Here there are paranormal encounters galore, but the biggest one is the Thunderbird.

A Thunderbird is a beast that stretches across the various folklores of many Native American nations. It is a supernatural creature that makes thunder when it flaps its wings, and it shoots lightning from its eyes. Some cryptozoologists have suggested the legend of the Thunderbird may have been passed down through oral history of interactions with the Teratorn, an enormous bird of prey that lived in North America up until the late Pleistocene era, about eleven thousand years ago.

Native Americans lived in North America for thousands of years by then, and would have lived in fear of the birds. An average Teratorn would have been about two and a half feet tall, with a wingspan of twelve feet. The Giant Teratorn was more than six feet tall with a wingspan of twenty-one feet. Other cryptozoologists say there may be residual pockets of Teratorns in secluded areas.

There may be one in Hockomock Swamp.

The swamp, a 16,950-acre freshwater bog, must not be too remote because there's a Wendy's Old-Fashioned Hamburgers close by. According to the *Yankee Express*, in 1988, two boys found enormous three-toed footprints and followed them into the swamp. The prints led to a bird, a bird of enormous size. When the bird became aware of them, it took to the sky.

When they returned home, panicked and crying, they told their families, their neighbors, and finally the police. The verdict? The boys had come face-to-beak with…a great blue heron.

Great blue herons are big, sure. Four and a half feet tall with a six-and-a-half-foot wingspan, but the boys knew what a great blue heron looked like, and that wasn't what they'd seen.

According to the *Express*, the bird, they said, was quite a bit larger than a heron, even taller than an average man, and it had human features.

Police Sgt. Thomas Downy from nearby Norton came forward and told his story.

In 1971, when Downy drove along the edge of the Hockomock wetlands and saw a bird. A bird six feet tall. His car spooked it, and it spread its wings—about twelve feet of them—and it shot into the sky. It did leave something behind—huge three-toed prints.

Michigan

Michigan is pretty far east to be in the Midwest, but there it is. It's also really, really wet. Bordering Lakes Superior, Michigan, Huron, and Erie, Michigan has the USA's longest freshwater coastline: 3,288 linear miles of it. In 1896, Henry Ford built his first automobile, the quadricycle, at his home workshop in Detroit. All that is cool, but not as cool as the fact that everyone's favorite B-movie star Bruce Campbell (you'll always be A-list in my book, Bruce) was born in Michigan. The state is also home to Dagganoenyent—the Flying Head.

Dagganoenyent

Legends of the Iroquois and Wyandot peoples of the Great Lakes Region feature a terrifying creature, the Dagganoenyent, otherwise known as the Flying Head.

This enormous flying head has long, stringy hair, blazing red eyes, and bat wings. It flies (hence the name) to chase down its favorite meal—humans. Like the Wendigo, the creation of a Dagganoenyent comes as a punishment for past cannibalism. Legend has it the Dagganoenyent hides in caves during the day, but emerges at night to feed. This head, at least six feet tall, brandishes claws to grab its prey, and, if it is attacked, wards off arrows and spears with its hideous, tangled hair.

If no humans are about at night, it terrorizes women in their homes, flapping against the walls, screaming at them to come outside.

Other stories have the Dagganoenyent as a spirit of vengeance. When the elders of the tribe refused a plan from the younger men on how to survive a famine, the younger men beheaded the elders and threw their heads into a lake. Unbeknownst to the mob, one of their own was dragged into the depths by the heads and drowned. The spirit of this man rose as a giant head to take his revenge on those who had killed him, and the village elders.

Minnesota

Although Minnesota prides itself as the Land of 10,000 Lakes, it actually has 11,842 lakes. Ten thousand is just easier to say. With all those lakes, Minnesota has 90,000 miles of shoreline, which is more than Florida, Hawaii, and California. Not each of them. *All* of them. Given all that water, there's one recreational boat for every six people in the state. Also, given all that water, you'd think the following monster would be in a lake. Nope. Minnesota has a dogman.

The Minnesota Dogman

In 1999, according to a story on KDHL in Faribault, Minnesota, an airman working late drove through a forested part of the base and saw eyeshine. The eyes were level with his—in his patrol truck. When the lights of the truck hit the creature, it turned and fled into the trees. The airman said it looked like a wolf, but was standing, and ran away on its hind legs.

The station reported another sighting from 2009 near Fergus Falls, a town of a little over fourteen thousand people. A man stopped on a rural road to watch deer on the roadside when he noticed a seven-foot-tall, bipedal wolf standing behind a tree as if it were stalking the deer. The deer noticed the car, then shot off the side of the road—and the wolf-man turned and glared at the driver in anger before loping off after the deer.

The station reported another sighting near Long Prairie, a town of fewer than four thousand people. A man hunting squirrels saw a human dog sitting near the woods, so, you betcha, he took a shot at it. The beast ran off and left behind only large canine prints.

Mississippi

Mississippi is in the South. Pretty obvious, right? It has 587 Dollar General stores, 88 Waffle Houses, and 26 places named Magnolia. The blues were born there, as were public colleges for women. The first human lung transplant was performed in the state capital of Jackson in 1963. Another first? The first Coca-Cola was bottled in Mississippi. But which was the greatest first? You'll have to decide. You'll also have to decide about the validity of the Chatawa Ape-Man.

MINNESOTA DOGMAN

Chatawa Ape-Man

The story is as old as trains and circuses. Most states have a story just like this.

A train carrying a circus to New Orleans derailed near Chatawa, Mississippi, in the Tangipahoa Swamp (okay, so the names change, but seriously, nearly every state).

When the train tipped, one of the circus's main attractions, a "half-ape, half-man hybrid" escaped, according to a story in the Hinds Community College newspaper.

The creature "was terribly ferocious and would attack any person or animal that got too close, so he had to be kept in a heavy iron cage in his own railcar."

The Ape-Man disappeared into the swamp, and, although teams of hunters were sent in pursuit, they never caught the beast.

Bigfoot sightings in Tangipahoa Swamp are to this day blamed on the Chatawa Ape-Man.

Missouri

My home state. It's kinda neat, I guess. Missouri is the home of the first bakery offering sliced bread (in Chillicothe), it has the City of Fountains (Kansas City), it's the birthplace of Mark Twain (Samuel L. Clemens, Florida) and Harry S. Truman (Lamar) and outlaw Jesse W. James (Kearney). It has the world's first drive-through restaurant, Red's Giant Hamburg, in Springfield (opening in 1947 and is *still open*). There's also a story about the Jimplicute.

Jimplicute

Deep in the Ozarks is a bloodlusting, furious monster called the Jimplicute that would often tear trees from their roots to satiate its anger. If that didn't satisfy the beast, the Jimplicute would rip up farmers' fences, screaming before it trudged back into the woods.

In the 1951 book *We Always Lie to Strangers: Tall Tales from the Ozarks* by Vance Randolph, the Jimplicute is described as an ancient ghostly dinosaur that lurks in the shadows of gravel roads, waiting to leap upon lonely travelers to suck their blood.

The creature is strange, seldom spoke of, and unlikely, but at least two newspapers claimed its name: Illmo, Missouri's *Jimplicute* (1960 to 1988), and Texas's fifth-oldest newspaper, Marion County's *Jimplecute*.

Montana

Montana has it all. The Great Plains, the Rocky Mountains, Glacier National Park, snow, snow, and snow. Not so much sun. Oh, wait; it reached 117 degrees Fahrenheit in Medicine Lake on July 5, 1937. So, I guess Montana does have it all. Except people. The population of the state is about 1.123 million. The populations of antelope, deer, and elk in Montana are each more than 1.123 million. But that's okay, wildlife's where it's at. Montana also has naked space aliens.

Naked Aliens

A trail camera caught something in 2022 that may be the cause of a future intergalactic incident.

According to WBZ News Radio, near Deer Lodge, Montana, Donald Bromley's camera took a photograph of a gray alien. Bromley told WBZ that area is a hotspot for UFOs. Lights in the sky, cars stalling, dead batteries, the works.

I'm not glossing over the naked thing.

"The more I look at it," Bromley told the station, "it was just odd. It was out of place and everything just matches the alien persona. The bigger bulbous head, you can tell he has no clothes, it's kind of a transparent being."

Patrick Cutler, who shot the documentary *Redgate* about space alien contact in Montana, told the station Bromley said he had more cameras in the area, but the one that took the photograph of the alien was dead, although all the others still worked. Because of its remoteness, Bromley didn't think anyone tampered with the camera, at least anyone human.

"To get to that spot publicly," Bromley told Cutler, "it's miles, literally miles to get there, and you probably have to drop 1,000 feet in elevation just to get to that point. It's unrealistic."

Nebraska

The Great Plains state of Nebraska is known for corn farming, the University of Nebraska Cornhuskers, and the early railroad, but Nebraska is home to other things, like more corn, Carhenge, investor Warren Buffett, even more corn, and being the birthplace of the Reuben sandwich, created in 1925 by Omaha grocer Bernard Schimmel. God bless you, Bernard. Nebraska also has a Bigfoot.

Oakland Creature

About an hour's drive north of Omaha, Nebraska, is Oakland, population 1,456, and on July 4, 1974, the lives of a local farm couple, Dale and Linda Jones, changed.

According to the *Flatwater Free Press*, the couple awoke in the middle of the night to a scream they thought was one of their hogs. Then their German shepherd began barking.

OAKLAND CREATURE

Dale went to look, but found nothing amiss. When he walked back toward his house, the scream again split the night.

"It was like something I'd never heard before," Jones told the *Free Press*. "It's indescribable."

He ran to the house and fetched a baseball bat and found his wife comforting their dog. Then he and Linda saw what had made the noise. It was built like a man, but too big to be one.

"It took off running on two legs, between the corncribs toward the cornfield," Linda told the newspaper.

They went inside and locked up the house, but couldn't get back to sleep. They were the first people to see the Oakland Creature.

A few days later, thirteen-year-old paperboy Nick Wickstrom delivered the *Omaha World-Herald*, his father and brother helping him on his route outside town, when they all saw something big running across the road.

"It looked like it was running on its front knuckles," Wickstrom told the *Free Press* in 2023. "It didn't look like anything I've ever seen before or since. There was no tail. Its legs looked like ours...I couldn't identify it. You know, we were country kids. We could identify just about every critter out there. But this one was different."

Other people came forth with stories of a six-foot-tall bear with a monkey face. Later, teenagers in a cemetery one night threw firecrackers at an upright, hulking, hairy figure. It ran off.

Although the local police tried tracking the beast, they never found it. Area residents put together creature-hunting parties, but came up blank.

By September, after Linda had delivered the Jones's first child, the sightings stopped. The identity of the Oakland Creature remains a mystery.

Nevada

Nevada is probably best known for Las Vegas, although it's much, much more than just gambling and white tigers. It's home to the Hoover Dam, the world's highest dam at completion in 1935; Death Valley, with the lowest point in North America, and the world record highest air temperature of 134 degrees Fahrenheit on July 10, 1913 (where did it set the record? Furnace Creek, of course); and the secretive Area 51. The less we say about that place, the better. Then there's the sad tale of Water Babies.

Water Babies

At 27 miles long, 11 miles wide, and 344 feet deep, Pyramid Lake is the biggest lake in Nevada. It should be. This saltwater lake is a remnant of an inland sea that covered Nevada in the Pleistocene era. It's called Pyramid Lake from a rock formation in the lake that looks like a pyramid. Its legend, however, doesn't reflect its beauty. Its legend is of the Water Babies.

To keep their tribe healthy and strong, the ancient Paiutes would take deformed and premature babies to the lake and drown them. Yikes.

After death, however, the babies of the lake refused to stay silent. Around dawn and dusk, people visiting the lake can hear the babies cry. As revenge for their deaths, the Water Babies are blamed for tragedies on the lake—anything from drownings, to boating incidents, to people disappearing without a trace.

Some of those people are fishermen. It is claimed the Water Babies lurk just under the surface of the water, waiting to snatch fishermen from their boats and drag them down into the depths.

Could the Water Babies be real? Fishermen have reported seeing strange objects on sonar that vanish as fast as they appear. So, who knows?

New Hampshire

New Hampshire is one of the smallest of the United States, but that hasn't hurt its popularity. This New England state is a haven for winter sports and is known for charming small towns and wilderness. Lots and lots of wilderness. The alarm clock was invented in New Hampshire, and the first US summer resort, Kingswood, opened there in 1763. The state is home to white-tailed deer, black bears, moose, and the Derry Fairy.

Derry Fairy

The sightings began in December 1956, when Alfred Horne, of Derry, ventured into the forest to find a suitable tree for Christmas, according to a report in the *Derry News*. As he chopped down an evergreen just the size to fit in his house, he saw a man watching him. The man was two feet tall, and, well, he was green.

Horne said it wasn't human. It had a pointed head, and ears like a sad dog; the skin was wrinkled like an elephant's, and the eyes looked like a snake's. Its arms and legs were too short for its body, and it had no toes.

They stared at each other for who knew how long before Horne pounced on the little green guy to cart him back to town, knowing no one would believe his story otherwise. The creature fought back and screamed. The scream was so loud and frightening, Horne gave up and ran off in terror.

According to *Weird New England: Your Guide to New England's Local Legends and Best Kept Secrets* by Joseph A. Citro, the story came from letters Horne wrote to Boston astronomer UFO researcher Walter H. Webb in 1962 and 1964.

New Jersey

New Jersey has 130 miles of northeast Atlantic coastline. It is the most densely populated state with 1,211.3 residents per square mile, but, despite that, 40 percent of the state is covered in forest. Cheerleading originated in New Jersey (Princeton in 1869), as did the drive-in movie theater (Camden in 1932). It's also home to a spoon museum with more than 5,400 spoons, and, of course, Hoppie, the Lake Hopatcong monster.

Hoppie

Lake Hopatcong is the largest freshwater lake in New Jersey at four square miles. It has a maximum depth of fifty-eight feet, but only an average of eighteen feet. Although it's in a state park, it's crowded with restaurants, a miniature golf course, an organic produce farm, lake cruises, and a water taxi.

However, in 1894, before all that when the lake was a quieter place, a fisherman claimed to have encountered a forty-foot-long creature in the lake. The August 4 edition of the *New York World* reported the incident.

The fisherman, A. Chamerlain, said he saw a snakelike creature with the head of a St. Bernard; its body was black, but its belly was dirty white. He wasn't alone. An unnamed local man claimed to have shot the creature in the head with a .38-caliber rifle, "And the bullet rolled off like water off a duck's back without even making the monster wink."

Over the years, explanations for Hoppie have included a large mud turtle, a beer keg, and, in 2000, a boa constrictor (although a boa's ten-foot length doesn't compare with Hoppie's forty).

There are eels in the lake, and loads of fish, although the biggest fish caught in the lake were by Norm Small who caught a 49-inch, 34.76-pound musky, and Alan Tuorinsky who reeled in a 28-inch, 12.9-pound brown trout, of Mount Arlington, New Jersey, both in April 2016, according to the *Bergen Record.*

At the time, Marty Kane, president of the Lake Hopatcong Historical Museum since 1990, told the *Record,* "Who knows what someone might have seen. Some people are scared when they see an eel, but they're a natural species in the lake."

New Mexico

New Mexico, the Land of Enchantment, sure is as advertised. There's the Trinity site (where the first atomic bomb was tested), Los Alamos National Laboratory (don't get me started on conspiracy theories), White Sands Missile Range, and the alleged 1947 UFO incident at Roswell (Demi Moore was born in Roswell and hasn't aged in forty years. No connection there). It also has a really weird monster called la Malogra. Weird, man.

La Malogra

The evil deed.

Throughout most of the year, la Malogra—the evil deed—does its evil through people (wicked strangers and awful neighbors), but when autumn rolls around, la Malogra rolls around.

When the seeds of the common Rio Grande cottonwood tree fall, the cottony balls that carry them along the wind form clumps that grow, and grow, and grow into a large ball that becomes possessed by la Malogra.

As this ball bounces along in the wind, it grows larger, and, at night, it hypnotizes anyone who sees it. The giant cotton ball then lunges at the person, and wraps them inside it, killing them. The next day, la Malogra's victim is left dead on the road.

New York

Unlike what the rest of the world thinks, New York isn't just a city, it's an entire state. Sure, New York City is there, and it has the Empire State Building (where that great cryptid Kong met his end), and the Statue of Liberty, but New York also has Niagara Falls, Buffalo (home of those tasty wings), and the Beamoc.

Beamoc

Willowemoc Creek, a tributary of Beaverkill River in Sullivan County, New York, is big with trout fishermen. There is one trout, however, no fisherman has ever caught, although some, like Paul Dahlie, have claimed to have come close.

Although *Beamoc* sounds like a Klingon word, it's the name for a two-headed fish that lives in Beaverkill River and its tributary, Willowemoc Creek, near Roscoe, New York. They take their fishing seriously there; the Catskill Fly Fishing Center and Museum is a seven-minute drive from town. In 2000, Dahlie had the hook of a "weighted pheasant-tail nymph fly."

Of course, the *Times Herald-Record* article published on April 1, so take the following report however you will.

BEAMOC

The story of Beamoc goes back more than fifty years. The trout supposedly lives in Junction Pool, the spot where the Beaverkill River and Willowemoc Creek meet. Legend has it the fish grew a second head because it couldn't decide which river to swim into.

When Dahlie (the newly appointed director of the Catskill Fly Fishing Center and Museum, according to the *Sullivan County Democrat*) cast his line on opening day in 2000 (April 1), he cast it into Junction Pool—at midnight. The tug on his line came seconds later.

He said it felt big, "But I had no idea how big," he told the *Times Herald-Record*.

As he fought with the fish, the line snapped, and a huge trout broke the surface. It had two heads; it was Beamoc. Dahlie told the newspaper he figured the fish's other head bit through the line.

Dahlie passed away March 13, 2004, so there's no confirmation of the Beamoc sighting.

North Carolina

North Carolina is an East Coast American state that's just north of, uh, South Carolina. As far as history can tell us, human beings settled in the area as early as ten thousand years ago. North Carolina is part of the Coastal Plain, the Appalachian Mountains, and the plateau region known as the Piedmont. The state is the birthplace of Pepsi-Cola, the place where Babe Ruth hit his first Major League Baseball home run (at the Cape Fear Fairgrounds in Fayetteville in 1914), and (drumroll, please) was the site of the first successful man-powered flight. It also has terrible names for its monsters, like Knobby.

Knobby

When people think of Casar, North Carolina, they...

Wait. Who's thinking of it? Casar is a town of 297 people in the North Carolina woods. It has two Baptist churches, a post office, and a Dollar General. There's also another Baptist church just outside town, and the South Mountain Bluegrass Church on the other side. South Mountain Bluegrass Church? I bet the hymns are incredible.

Originally called Race Path, in 1903, the locals wanted to change the town name to Caesar, but the name was misspelled on the incorporation application, and there you go.

Given its location, it's a great spot for monsters.

In the late 1970s, people all over Cleveland County, North Carolina, reported encounters with a big, hairy, humanlike entity the locals would come to call Knobby, named after an early sighting of a Bigfoot near Carpenter's Knob just off NC-10. The legend of the beast stretches back decades before that, but Knobby gained national attention in 2010 when Tim Peeler called 911 at about 3:00 a.m.

PEELER: "If I had a camera, I'd get a picture of it."

911 DISPATCHER: "A picture of who."

PEELER: "I don't know what it was. It was walking upright like a man. I would not kill it because I was afraid to."

The creature approached him before walking past him, across his yard, and onto a lane that went into the woods, he told WBTV in Charlotte, North Carolina.

However, Peeler didn't just watch Knobby as the beast came near, he "talked rough" to it, then poked it with a stick. Because I guess that's what you do when an eight-foot-tall humanoid ape trespasses on your property.

Peeler, fortunately, lived to tell his story.

Is it just a story?

A local, Sean Moses, told the nearby *Shelby Star* the story of when Knobby came about when moonshiners wanted to keep the law away from their stills on Dirty Ankle Road.

Others claim it's a prank.

Regardless, Knobby has become a fixture in Cleveland County. Area residents voted to name the Cleveland Community College's mascot the Yetis in honor of Knobby.

North Dakota

North Dakota is up there, way up there, and bordering two Canadian provinces, it's probably really polite. It exports tons of manufactured products, along with petroleum and coal. The clothes dryer was invented in North Dakota, as was the very first Kodak camera. The geography of the state is mostly the Great Plains and the Badlands. There's also a mermaid.

The Lake Sakakawea Mermaid

The third-largest reservoir in the United States, Lake Sakakawea stretches more than 382,000 acres, its depths reaching 180 feet. In that lake live some huge fish, paddlefish (seven feet long), and giant sturgeon (sixteen feet long). It's also supposedly home to a mermaid.

Local legends tell of a mermaid who, with her mate, swam the entire length of the Missouri River. However, when the US Army Corps of Engineers built the Garrison Dam in 1956, they built it between the lovers, trapping the mermaid in the lake.

The mermaid, however, isn't bitter toward humans, who imprisoned her, keeping her from her lover. People have reported hearing her singing at night and seeing a light in the lake, believing the mermaid is blessing the creatures who live with her in her prison. During the day, she stays low, in the deepest parts of the lake, awaiting the day she will be reunited with her mate.

In 1989, the city of Riverdale on the lake commissioned artist Tom Neary to sculpt a statue of the mermaid (they named her Misty). This two-hundred-pound, seven-foot-tall statue is in the city's central plaza, the base covered in rocks from each county in the state.

Ohio

With plains, hills, river valleys, and beaches of Lake Erie (water temps in the upper 60s and lower 70s in the summer. Not bad), Ohio is as pleasant as you can get with a Midwest state. It's home to two aviation firsts: Dayton is the boyhood home of the Wright brothers (who successfully flew the first airplane), and Wapakoneta is the boyhood home of Neil Armstrong (the first human to set foot on the Moon). The Rock 'n' Roll Hall of Fame is located in Ohio, as is the Crosswick Monster.

Crosswick Monster

In the early 1800s, two boys played near a creek when an enormous snake shot from the water. Arms grew from the trunk of the beast and grabbed one of the boys, pulling him through the forest and disappearing into a hole beneath a sycamore tree. Fortunately for the boy, it dropped him before entering the hole.

The boys ran to nearby Crosswick and told their story. More than twenty townsmen grabbed guns and axes and stormed the forest, chopping down the tree. The creature emerged from its den and loomed over the party. Seeing that it was outnumbered, the monster fled, destroying a fence and disappearing into a distant cave.

According to the *Columbus Navigator*, "It is described as being thirty to forty feet long, twelve to fourteen feet tall when erect, sixteen inches in diameter, and legs four feet long. It is covered with scales like a lizard's, of black and white color with large yellow spots. Its head is about sixteen inches wide, with a long forked tongue, and the mouth inside is deep red."

No one ever saw the Crosswick Monster again, but it made its mark in the monstrous legends of Ohio.

Oklahoma

Oklahoma produces natural gas, oil, and, for some reason, lots and lots of iodine. The state has dry plains, but it also has forests, is part of four mountain ranges, and has more man-made lakes (two hundred) than any other state. The official state meal (adopted in 1988) consists of chicken-fried steak, barbecued pork, fried okra, squash, cornbread, grits, corn, biscuits with sausage gravy, black-eyed peas, strawberries, and pecan pie. Oklahomans must be hungry. The state also has a nasty critter called the Stikini.

Stikini

According to Seminole folklore, Stikini are wicked witches that can turn themselves into a strange conglomeration of human and owl (similar to a story in Mexico. Hmm. Is there some truth to this?).

Stikini appear as humans during the day. Heck, you might run into one at Walmart. But, at night, they become undead owl monsters after vomiting up their souls and internal organs. Their food of choice? Beating human hearts—especially one from a child.

If a Seminole speaks the name of a Stikini, there is a chance they may turn into one.

The Stikini prefers forests to plains or cities, so venturing into the woods in Oklahoma at night is discouraged. But if you do, and the hooting of an owl grows closer, and closer, it probably wouldn't hurt to run.

Oregon

Considering the state of Oregon is so far up the West Coast of the United States it's practically Washington (which is essentially Canada), it's odd the Oregon coast was discovered by Europeans in 1592 when Greek sailor Juan de Fuca, working for Spain, floated by and said, "Hey, that's cool." (Artistic license, but I'm sure it was something like that). The view of Oregon gives us valleys, mountains, forests, and the Timberline Lodge, which was the scenic view of the Overlook Hotel in the movie *The Shining* (1980). This all makes Oregon a beautiful, and spooky, state.

Gumberoo

The thick forests of Oregon were havens for nineteenth-century lumberjacks. However, the bearded, flannel-wearing mountain men weren't prepared for what they encountered among the trees.

A black, hairless, fat, black-skinned, bearlike creature—the Gumberoo—terrorized them. Bullets couldn't penetrate its hide, so they tried anything they could to kill it, and found fire; when these beasts caught fire, they exploded, as cryptids sometimes do.

The beasts were hungry. They were fast and could pounce upon a cow or horse in a flash, devouring the entire beast before moving on.

Of course, the lumberjacks blamed forest fires on the Gumberoo. Anytime a fire broke out, they'd claim to hear an explosion in the trees. The fires had nothing to do with their campfires, or cigarettes, or doing stupid man stuff. It was because of the monster.

To commemorate the Gumberoo, Rogue Ales from Newport, Oregon, brews Gumberoo, an American IPA that checks in at 6.8% ABV. Oh, wow. Chug-a-lug.

Amhuluk

This monster is a movie star. Sort of.

The Amhuluk is a horned water dragon from the traditions of the Kalapuya of the Willamette Valley between Portland and Eugene. The dragon, which lives in a lake near Forked Butte, is covered in spotted fur, all except its legs, which are hairless. Its shtick? Oh, the usual for dragons: death, disease, drowning, and carrying with it a sickness-bearing fog. The Amhuluk is mentioned in the movie *Godzilla: King of the Monsters* (2019) but doesn't appear in full until featured in the novel *Godzilla Dominion* (2021).

Pretty cool for a monster most of us haven't heard of.

This dragon is its own personal junk drawer, keeping items it might need later tied to its spotted body. When it travels around its lake, dogs it had tamed (and are equally spotted) travel with it, like a witch's familiar. As it moves around its lake, it drowns every living thing it can find: people, deer, the sky—and bears.

Bears are attracted to the Amhuluk's lake (mostly because the dragon enchants them), where they encounter a monster called the Atúnkai (more on that soon).

A Kalapuya story pits three siblings against the Amhuluk when they ventured too close to the dragon's lake. The children wanted to vanquish the monster and take its horns back to their village,

AMHULUK

but the dragon used its horns to impale two of the siblings. The third ran home and told his parents of the deaths. The father ran to the lake and began a five-day ordeal to rescue his children, but in the end, he went home, because the Amhuluk had taken them.

Now, let's get back to the bears.

Atúnkai

The Atúnkai and Amhuluk live together in a lake on Forked Butte.

Good for them. I hope they're happy.

When the Amhuluk enchants large animals—bears, and sometimes people at the end of their lives—it uses its magical dragon powers to change them into a different creature, like sea otters, seals, or Gonzo from the Muppets. (I just lied about Gonzo from the Muppets.)

The change turns these beings into the Atúnkai. The Atúnkai live in this lake, and neighboring lakes, and behave like sea otters and seals, which is a playful end for the bears and people in their old age.

A great retirement plan.

Pennsylvania

The Keystone State (apparently not named after the cheap Molson Coors Beverage Company beer. Who knew?) played an important part of the formation of the United States. Independence Hall in the city of Philadelphia is the location of the signing of the Declaration of Independence (and the location of a great Nic Cage movie). It's not free of monsters, however. It has lots. Like the Giwoggle.

Giwoggle

Thanks to local Clinton County historian Lou Bernard, as of 2011, the Giwoggle is the official monster of Clinton County, Pennsylvania.

A Giwoggle, unlike its name, isn't silly. Not at all. This bipedal wolf looks to be much like a dogman or werewolf, although it has claws instead of humanlike hands, and hooves for feet. This is probably because it's not natural—it was created by a witch.

In 1870, on Keating Mountain in Clinton County, a coven of witches summoned otherworldly beasts that resembled wolves that walk on two legs. Reports of these Giwoggles have continued in the county since.

Rhode Island

The New England state of Rhode Island is, of course, not an island. However, no one can agree why it's called one. One story is that Dutch explorer Adrian Block called it *Roodt Eylandt*, which meant "red island," or "root eyelid," or "rude assbutt." It could be anything, because apparently spelling wasn't a thing back then. Another story has Italian explorer Giovanni da Verrazzano commenting that an island in the bay looked like the Greek island of Rhodes, and the name stuck. Regardless, it has a werewolf.

Pawtucket Werewolf

When a group of students decided to take a walk in the forest after completing their final exams, they soon decided they'd made a serious mistake.

First, it was December 16, 2008, in Pawtucket, Rhode Island, which isn't known for balmy winter weather.

Second, they met a werewolf.

As they walked farther and farther into the woods, the forest sounds fell silent. A huffing breath. A twig snap. They realized they were not alone. When they saw their unwelcome visitor, it was more than six feet tall, with the body of a man and the head of a wolf.

The beast froze, and looked around, apparently aware of the students, but not knowing where they were. Suspicious of the encounter, the wolf-man turned and ran deeper into the forest, away from the students.

When they were sure it was gone, they ran. *Aaaa-oooo.* Werewolves of Pawtucket. Kinda rolls off the tongue, doesn't it?

South Carolina

The South Carolina state motto, "Like North Carolina, only South" (totally not the state motto). With ocean beaches, islands, plantations, Fort Sumter, Myrtle Beach, Charleston, and humidity, South Carolina is a beautiful state with 187 miles of Atlantic coastland, the Blue Ridge and Piedmont Mountains, and winters I'd prefer over all the damn snow. And, yep, it has the Gray Man.

The Gray Man

Pawleys Island is the oldest resort town in South Carolina. The island (is it an island?) is separated from the mainland by a salt marsh and is four miles of beaches and sand dunes.

The Gray Man is just a pest.

In the 1820s, people on Pawleys Island began to see a cloaked man walking the beaches; soon after, a major storm struck the island. In 1954, a woman said she saw him on the beach right before Hurricane Hazel hit. Someone else reported seeing the Gray Man in 1989 before Hurricane Hugo, another still in 2018 preceding Hurricane Florence.

But who is the Gray Man?

As legend has it, a man traveling home to see his fiancé fell victim to a spooked horse and the salt marsh. The horse became stuck in the marsh and dragged the man beneath the brackish water to his death. The fiancé, as if looking for her dead love, often walked the beach near where he died. One day, she saw him, and he told her a storm was coming, then he vanished.

South Dakota

South Dakota, home of rolling prairies, Wall Drug (it's a flashy tourist trap, but still a working drugstore), the starkly beautiful Badlands, the Black Hills National Forest, and the monuments Mount Rushmore and the Crazy Horse Memorial. Near those monuments is Rapid City, the City of Presidents, with life-size statues of presidents on the sidewalks. Check it out. Also check out one of South Dakota's cryptids, the Thunder Horse. It's thunderific.

Thunder Horse

The Santee Dakota tribe had a legend of an enormous horse whose hooves cause thunder.

The Thunder Horse was an ancient beast that left the great rolling plains of the Dakotas before time, but it hadn't left forever. Occasionally, the noble horse would come down from the heavens to run, and when its hooves struck the ground, thunder rolled in the sky.

The horse would chase bison and antelope over the prairie, its hooves creating sparks that shot into the sky. When that happened, rain fell, and, exhausted, so did the Thunder Horse, its bones turning to stone. Hunters, finding the bones, took them back to their village, and told stories of the Thunder Horse.

Wait. There's more.

When American paleontologist Othniel Marsh visited the area in the late 1800s, he requested the locals take him to the spot where the Thunder Horse bones were found. When he got there, he and his team dug and discovered the mostly complete skeleton of a Megacerops, a creature from the Eocene epoch related to a horse, but that looked like a rhinoceros. He determined that from this fossil the story arose.

Tennessee

A Southern state, Tennessee is full of country music and food. Nashville, the state capital, hosts the Country Music Hall of Fame and the long-running Grand Ole Opry. In the southwest corner is Memphis, once home to Elvis Presley, and still home to his mansion, Graceland. And, whenever you make it to Memphis, make sure to try the barbecue. Tennessee is also home to a creature called the Not-Deer.

Not-Deer

You'll know something's off with the deer at first glance. The proportions are all wrong. It has too many joints. It's big, like an elk. Oh, wait. No. It's the eyes. The eyes are wrong; they face forward, and the head is like a cow's. None of this is right. Oh, man. It stood up, on its hind legs. It's coming closer!

The Not-Deer is an Appalachian cryptid mostly sighted in Tennessee. Although not normally aggressive with humans (but definitely not afraid of us), this intelligent creature will attack if threatened. An attack from this strange beast is frightening enough, but what makes it even more frightening is witnesses have seen Not-Deer eating meat.

NOT-DEER

What could a Not-Deer be? A deformed animal? A deer with chronic wasting disease? An alien creature? A beast of Native American folklore? A completely unknown animal?

Whatever the case, the Not-Deer sounds freaky.

Texas

Texas is big. Really big. As its 1995 slogan bragged, "Like a whole other country." It was, once upon a time. In 1836 until 1846 it was the Republic of Texas. During its existence as a European-settled entity (let's not forget the Paleo-Indians), Texas was under the flags of Spain, France, Mexico, the Republic of Texas, the Confederate States of America, and the United States. I think it's still under the flag of the last one, but you never know. Texas has a total area of 268,820 square miles, and a population of 29 million people. Its economy is equal to India's. Wow. Remember the Alamo, Bluebells, football, cowboys, and armadillos dead on the highway, Texas is a beautiful state, and is the home of the Bear King of Marble Falls.

The Bear King

An area just northwest of Austin, that is now Colorado Bend State Park, was once the home of the Kickapoo peoples. When white men moved into the area, the Kickapoo warned them of the Bear King that ruled that land. Of course, when it comes to the legends and beliefs of Indigenous peoples, we usually don't pay attention.

One of the first white victims of the Bear King was Ramie Arland, who was outside tending to the family farm animals when she was assaulted by a bear man that grabbed her and spirited her into the wilderness, according to a 1901 edition of the *San Francisco Chronicle*. A day later, a local man found Arland in the forest; she told him the bear man took her to a cave. Arland waited until the creature was asleep, then escaped.

The *Corpus Christi Caller Times* reported a group of hunters armed themselves and marched through the woods to the cave. A manlike bear emerged from the cave beating its chest like a gorilla, and growling like a mountain lion—but no one wanted to shoot it because, despite the thick pelt of fur, it looked too human.

Then it charged, and the hunters shot the Bear King dead.

However, the memory of the Bear King survives in the small town of Marble Falls in the form of the Bear King Brewing Company.

"A mysterious creature stalked the outskirts of Marble Falls," the company wrote on its website. "Some say it was half-man, half-bear."

Now, the Bear King is 100 percent beer.

Beast of Bear Creek

Cleo, Texas, doesn't exist. Not really. At least not anymore.

Kimble County, Texas, about 150 miles west of Austin, was created in 1858, and is named after George C. Kimble, the commander of the Immortal 32, a force from the Texan Militia's Gonzales Ranger Company who fought and died during the battle of the Alamo in 1836. The population of Kimble County is a little more than 4,000 people.

The county's most famous town, Cleo, founded in 1880, is nothing but a ghost town now, all but three of its remaining citizens moving out by 1990. But, this spot in the road, in South Llano River State Park, is known for a monster—the Beast of Bear Creek.

When the town was first formed, its settlers were faced with a problem; they wiped out the local Native tribes, leaving one survivor—a shaman.

The massacre, of course, is the problem, and problems beg for solutions. This solution was the shaman, who had the ability to transform into a wolf at night, a wolf of vengeance, who ripped the throats from everyone who killed his friends and family.

Over the years, people have speculated the Beast of Bear Creek could be a Bigfoot, black bear, mountain lion, werewolf, or skin-walker. Of course, it could be just what the legend says: a wizard taking revenge for the senseless slaughter of his people.

Mystery Animal of Rio Grande Valley

Mission (population 87,000) in the metropolitan area of the cities of McAllen and Edinburg, Texas, and Reynosa, Mexico. In 2023, it was also home to a mystery animal. Rangers in Bentsen–Rio Grande Valley State Park saw game camera footage of a creature they couldn't identify.

Texas Parks and Wildlife Department posted a gray picture of the animal on its social media asking viewers, "Is it a new species? An escapee from a nearby zoo? Or just a park ranger in disguise?"

While the third question was all in fun, what was the animal? It was a long quadruped with short ears and a plump body. It was difficult to identify the tail.

Speculation ranged from the jaguarundi—a wildcat that lives from northern Mexico to central Argentina that is slightly bigger than a domesticated house cat—to a common American badger, although the creature in the photograph didn't possess the usual markings of a badger.

Its exact identification is still unknown.

Horizon City Monster

Horizon City, with a population of 22,489, sits inside the city of El Paso (population 678,815) in the northernmost part of the Chihuahuan Desert. Although mostly bordered by the city limits of El Paso, it's also connected to the Hueco Mountains to the east.

In 2001, according to the *El Paso Times*, something came down from those mountains and terrorized a resident of Horizon City.

Retired secretary Cecelia Montañez was at the edge of the desert when she encountered a tall, furry, humanlike creature, nearly eight feet tall, with glowing red eyes and short maroon fur. "I saw a big gorilla-like thing walking toward the desert," Montañez told the *Times*.

According to the *Times*, a number of residents of Horizon City have reported seeing the beast near Lake El Paso, or in the desert.

But, the 2001 incident wasn't the first. The *Times* ran an article on September 20, 1975, about three teenagers hanging around the Horizon Golf Club who saw a gorilla. The El Paso County Sheriff's Department deputy who investigated the case, Bill Rutherford, told the *Times* in 1975 he "didn't think it was for real." Although he retired in 1988, he was around in 2003 when the *Times* revisited the Montañez story, and never changed his opinion. "I was a deputy sheriff, so I was regulated to do something," Rutherford told the *Times* in 2003. "I never saw it. I thought it was a hoax."

Utah

The state of Utah is a bit of a strange place. With pine-forested mountain valleys and arid deserts (Utah is the second-driest state in the union, just after Nevada), it has sand dunes, and ski resorts, the Great Salt Lake, and forty state parks (it's the only state where every county has at least part of a state park). It's part of the Four Corners region where the corners of four states touch (apart from Utah, it's Colorado, Arizona, and New Mexico). One of its state symbols is a Dutch oven. It's the leading manufacturer of (tractors? Airplanes? Life-saving medicine?) rubber chickens. It also has a lake monster named...

Old Briney

On July 8, 1877, a group of workers for Barnes and Co. Salt Boilers near the shore of the Great Salt Lake heard something strange over the water. When they got closer to get a better look, a monster, "a huge mass of hide and fin rapidly approaching, and when within a few yards of the shore it raised its enormous head and uttered a terrible bellow," according to the July 1877 *Salt Lake Herald-Republican*.

One of the men, J. H. McNeil, said the monster was "a great animal like a crocodile or alligator, approaching the bank, but much larger than I had ever heard of one being. It must have been seventy-five feet long." The beast's head, however, looked more like that of a horse. "When within a few yards of the shore it made a loud noise and my companion and I fled up the mountain, where we stayed all night. When we came down in the morning we saw tracks on the shore, but nothing else."

Vermont

Vermont is a postcard. Covered by mountains and forests, Seventy-eight percent of Vermont is covered in trees, and is an autumn foliage fan's dream. On the subject of trees, it's the United States's top maple syrup producer. Apart from the landscape, it is home to more than one hundred covered wooden bridges from the nineteenth century, and is also home to Ben & Jerry's Ice Cream and vampires.

The Vampire of Manchester

Rachel Harris Burton, twenty, married Captain Isaac Burton of Manchester, Vermont, in 1789. By all accounts, she was a healthy, beautiful woman. However, she quickly became ill, and her health quickly began to fade. In 1790, Rachel Harris Burton died; Captain Burton buried her in the city's Factory Point Cemetery.

By the next year, Captain Burton had already remarried, this time to a woman named Hulda Powell, who began to suffer from the same illness as Rachel. To the townsfolk, there was only one possible explanation to the sickness that claimed Rachel and was killing Hulda—a vampire.

To cure Hulda, the locals believed exhuming Rachel and retrieving her internal organs would do it. Of course, they'd have to burn them first. In February 1792, they dug up the captain's first wife in hopes of using her to drive the demon from his second. After disinterment, they removed her liver, lungs, and heart, and burned them in the forge of the town's blacksmith.

It didn't work. Hulda Burton died in 1793.

A vampire panic had spread across New England during the late 1700s and early 1800s. Strangely enough, so did tuberculosis, also called consumption because the illness appeared to consume its victims, sucking their vitality and causing them to appear thin, weak, and pale, as if, well, as if a vampire were draining their blood.

Science discovered the cause of tuberculosis in 1882; the cure wasn't discovered until 1943, and it wasn't the smoke from the burning internal organs of a vampire. Strangely enough, the cure was medicine.

Virginia

Virginia, one of the original thirteen colonies, holds a lot of early United States history, such as our first president George Washington's home, Mount Vernon, and third president Thomas Jefferson's plantation, Monticello. It has farmland, mountains, and Atlantic coastland. And, Devil Monkeys.

The Devil Monkey

Devil Monkeys have been reported in multiple states, including in southern Virginia. The descriptions are all the same. These primates resemble baboons and come with a bad attitude. They are strong, fast, agile, and always angry.

Much like the Chupacabra, Devil Monkeys seem to have a taste for small livestock, but unlike the Chupacabra, they aren't relegated to vampirism. They shred the animals and eat what they want.

Baboons, native to Africa, are the largest nonhuman/non–great ape primate growing up to three feet long, five feet counting the tail, and weighing up to eighty pounds. As far as science is concerned, baboons (that have existed for about two million years) have never lived in North America. The last primate in North America was the Ekgmowechashala, which existed thirty million years ago and were about the size of a lemur.

Extant Ekgmowechashalas? Not likely.

Then what are they? Escaped pets? New species? Ugly, poorly behaved children?

Much like having an answer for all the monsters in this book, we'd have to capture one to make sure.

Washington

Nestled in the top west corner of the contiguous United States, the State of Washington doesn't cause much trouble. It's covered in trees, lakes, mountains, glaciers, and hipsters, who enjoy nature and the state's enormous number of wineries (more than 1,050), and breweries (426, at last count). Washington has 3,000 miles of coastline, thick rainforests, bays and fjords, and sixty-four mountain ranges. Washington is the birthplace of musician Jimi Hendrix, Starbucks coffee, Boeing, Amazon, and dragons?

The Dragon of Lake Chelan

With an area of 33,344 acres, Lake Chelan in Chelan County, is large, and deep. Its maximum depth is 1,486 feet, making it the third-deepest lake in the country. That's plenty of room for a monster to swim.

And swim it has, at least since 1892. That year, the *Wilbur Register* reported a white creature with "the legs and body of an alligator, and the head and restless eyes of a serpent" erupted from the water and bit one of the men. The monster had bat-like wings and was covered with scales.

The other two men attempted to set the beast on fire, but it dove, taking the attacked man into the depths, according to a story on kissfm1053.com.

Seattle's NBC television affiliate, KING5, interviewed a storyteller with the Chelan Museum who told the station people have reported unusually large waves overturning their boats, and some people who've claimed to actually see the beast. Men standing on a boat dock in 1910 reported an encounter with a seventy-five-foot-long monster surface nearby.

A school bus crashed into the lake in November 1945; unfortunately the accident killed fifteen children and the bus driver. Legend has it divers who went to locate the bus so it could be extracted from the lake were chased by a monster about seventy-five feet long; however, news reports at the time mention nothing about the divers seeing the creature.

Apparently, some people call the dragon Ogopogo, although that particular lake monster lives around 220 miles away in Okanagan Lake in British Columbia, Canada. So, get your stories straight, eh?

West Virginia

West Virginia, one of only two states added to America during the Civil War (also, Nevada), is small. It's the tenth smallest by size, and has the twelfth-smallest population, although that doesn't mean it's not big in other areas. It ranks third in most forested states, which is expected considering the entire state is within the Appalachian Mountains. But, it's most famous for monsters. Doesn't West Virginia get enough attention with Mothman, Goatman, and the Flatwoods Monster, without pulling something creepier into the mix? Tell that to the Grafton Monster.

Grafton Monster

Robert Cockrell drove fast. It was nearing 11:00 p.m., the night of June 16, 1964, as the young newspaper reporter sped home from work at the *Grafton Sentinel*, and he just wanted to get to bed.

As he negotiated a curve on Riverside Drive, he saw an enormous bipedal creature on the side of the road. The beast stood between seven and nine feet tall, its shoulders wider than any linebacker's.

A Bigfoot?

GRAFTON MONSTER

Maybe. A bald one.

Cockrell's headlights moved across a blindingly white, hairless, humanoid entity, its shoulders so big and muscled its head wasn't apparent. As the young journalist cruised by, the monster walked across the road behind his car.

According to Cockrell's 2022 obituary in the *Dominion Post* in Morgantown (about twenty-five miles north of Grafton), he became the first person to report seeing the Grafton Monster, later known as the Ogre, or, even better, the Grant Town Goon.

Even though Cockrell was terrified (and who wouldn't be?), instead of going home to bed like he wanted, he hunted down a couple of friends, and went back to the spot he saw the Goon. Even though the only physical evidence they found was trampled weeds and grass, they were followed by a creepy, hair-raising whistle as they explored the area. According to Bigfoot researchers, this whistle is common during encounters.

The *Sentinel* published a story on Cockrell's encounter, which spurred locals to go out and find the monster. Two days later, June 18, a *Sentinel* headline read: "Teen-Age Monster Hunting Parties Latest Activity on Grafton Scene." The article said groups of teenagers were out looking for a monster "nine feet tall and about four feet wide." Groups of teens drove up and down Riverside Drive, some of them claiming to have seen the monster, although one of them speculated it was a polar bear.

Wait? A polar bear? In West Virginia?

To keep the traffic down (and the increased possibility of a dangerous incident on the crowded road), local law enforcement encouraged the *Sentinel* to bury the monster story. The excitement eventually died down, but every once in a while, someone claims to have encountered the Grafton Monster.

While writing this, the town of Grafton is making the final preparations for its Grafton Monster Festival planned for June 15 and 16, 2024. Planned are speakers, a cryptid cosplay contest (if I can make it, I'm totally going as a Devil Monkey, or, *ooh-ooh*, a Jackalope), a June 16 Father's Day cornhole tournament (if you don't know what the game cornhole is, it's not as awful as it sounds), an art exhibit, and, of course, vendors.

I hope your festival rocked, Grafton.

Wisconsin

Farms, forests, lakes, beer, cheese... Wisconsin doesn't need much else. The thirtieth state touches two of the Great Lakes, Michigan and Superior, and is known for the Green Bay Packers, cheesehеads, the birthplace of Harley-Davidson, and cool city names like Oconomowoc, Sheboygan, and Kenosha. Then, there's the Devil's Lake Monster in another cool name, Baraboo.

Devil's Lake Monster

The name *Devil's Lake* is pretty ominous. The name the Native Nakota Sioux used for the lake wasn't much better, *M'de Wakan*— "Bad Spirit Lake."

It apparently gained its name honestly. A tentacled monster was known to lurk in the lake, often grabbing and pulling hunters and fishermen to their watery deaths. Stories change over the years, and the monster of Devil's Lake is now said to resemble a plesiosaur.

Legends brought people to the lake during the 1920s through the 1930s, although now they mostly come for the fun. The frigid Devil's Lake is about a mile long and half a mile wide with a maximum depth of forty-seven feet, but averaging thirty feet. The lake and surrounding parkland offers thirty miles of hiking trails, beaches, biking, hunting, and fishing.

Wyoming

This sparsely populated state (seriously. The population of Wyoming is 581,381 people. The city of Memphis, Tennessee, has more) is known for its beautiful landscapes, Yellowstone National Park, and rodeos—Cheyenne Frontier Days is called Daddy of 'Em All. There are also creepy cryptids.

Casper Mountain Crawler

Two young men hiking to the Garden Creek Falls area of Casper Mountain (that's just south of, surprise, surprise, the city of Casper) discovered something that made them wish they hadn't gone out that evening in 2022—the Casper Mountain Crawler.

While taking a video to post on Snapchat, one of the young men, only going by the name Isaiah, said they heard something large approach. When it got close enough, it growled at them. The video shows a grayish-white humanlike figure lurking in the trees. Then the video ends.

Isaiah said in a later video on TikTok that Snapchat cut their time, and they ran.

"Whenever I recorded, whatever I recorded, I wasn't going to sit there and record it because I was running for my life," Isaiah said. "I'm not trying to be funny about this. Literally, I was puking."

Chapter 12
The United Mexican States

Mexico has been the home to civilizations with permanent settlements, advanced agriculture, science, and medicine for thousands of years. The Olmecs were the first, beginning at around 1600 BCE, then were followed by the Teotihuacan, Toltec, Zapotec, Maya, Aztec, Mixtec, and Chichimec. Today, Mexico excels in education, health care, and revels in music and cuisine. It is the fourth-largest country in North America behind Canada, the United States, and Greenland, and with 761,600 square miles of land, it's the thirteenth-largest country in the world. Its capital, Mexico City, is the fifth-largest city on the planet with a population of 21,671,908. With 5,800 miles of coastline, mountains, volcanos (the Pico de Orizaba volcano, at 18,491 feet, is the third-highest peak in North America), tropical rainforests, deserts, and the second-largest coral reef on Earth, Mexico is a nature lover's paradise. And, for history lovers, Mexico City has the second most museums in the world. Mexico also gave us Ricardo Montalbán and Salma Hayek, so, gracias. In a country this large, with thirty-one states and one federal district, Mexico is rife with monsters.

Aguascalientes

The Mexican state Aguascalientes (Hot Waters) is named such because of its abundance of thermal springs and is filled with, of course, spas and health resorts. Located in central Mexico, Aguascalientes, is part of the Bajío region in the central Mexican plateau. It also has lots of silver that, for some reason in the 1600s, the Spanish kept coming back for. Aguascalientes is known for the San Marcos Fair (a 190-year-old national fair) and the National Museum of Death. Um, okay. I'm going to the fair.

Giants

Enormous humans have a long history in Mexico, and I'm not just talking luchador Alberto Del Rio (seriously. The wrestler is 6'5" and 239 pounds of pure muscle). According to Aztec legends, the giants Cuauhtemoc, Izcoalt, Izcaqlli, and Tenexuche held up the sky. Other giants, like Mixtecatl, Otomitl, and Xlhua founded cities and built pyramids. These giants weren't just stories, they were part of Aztec history.

The stories of some giants, though, have continued into modern times. In 2023, a truck driver in Aguascalientes witnessed something he couldn't believe.

Driving through a hilly region between the state's capital city (also called Aguascalientes) and Lagos de Moreno, the trucker saw a humanlike creature walking along a ridgeline. He stopped and took a video. It showed an enormous man on the summit.

Of course, the truck driver put the video on the most trustworthy website available—TikTok.

That posting doesn't discount the sighting.

The Chichimeca peoples that inhabited the region in the twelfth and thirteenth centuries also knew giants. One of them, Nahuatlaca, was a friend. Nahuatlaca saved the Chichimeca from being wiped out by neighboring tribes, but once that was over, the giant became their leader; he also became a monster.

You don't pay taxes? Death by club.

You don't feed him enough? Death by club.

You don't look at him the right way? Death by club.

The Chichimeca plotted against the giant and, you guessed it, death by club.

Nahuatlaca implored the gods to protect him, which worked—for a while. The Chichimecas finally were freed of this wicked savior when they found him sleeping and dropped a huge basket of rocks on his head, burying him. They still worry he may one day free himself from his stony grave.

Chan del Agua

The largest lizard in Aguascalientes is the giant horned lizard. Wow. Giant horned lizard? How big can a giant horned lizard get? Let's see. Oh, 8 inches long and 3.5 ounces. Well, the giant horned lizard isn't really the largest lizard in Aguascalientes anyway, because it has nothing on Chan del Agua.

A pond named el Campanero (the Bell Ringer) formed where Río San Pedro (Saint Peter River) meets Paso de Curtidores (Tanners Pass). The ancient pond, at the intersection of Río San Pedro

CHAN DEL AGUA

and the street Salvador Quezada Limón (which means, in Spanish, "Salvador Quezada Limón"), was, in the 1800s, used by soldiers to wash their horses. The pond was also home to something big and green—Chan del Agua.

The legend of the Chan del Agua is an ancient one, dating back before the Spanish conquest of Mexico. He is a lizard the size, and shape, of a man, possessing a tail and shiny green scales. The Chan is protective of water and defends it, killing those who attempt to foul it and choking to death those who even speak ill of it (who speaks ill of water?).

He was also a randy fellow.

For a time, when women went to bathe in the pond, they became pregnant. I bet that conversation was uncomfortable:

Husband: "You're what?"

Wife: "Pregnant."

Husband: "But, how? I just returned from a four-month journey."

Wife: "I made a mistake! I bathed in el Campanero!"

Husband: "Chan del Agua?"

Wife: "Yes."

Husband: "*Again*?"

Of course, when the wave of babies grew to resemble the local soldiers (the ones who bathed their horses in the pond), Chan del Agua didn't get all the blame, although, honestly, he still got some.

Not to let any good monster go unexploited, el Chan Del Agua is a pub in nearby Jalisco, Mexico, that has 4.5/5 stars on Restaurant Guru, with reviews like this:

"Muy buen ambiente cerveza bien frías la botana excelente" (Very good atmosphere, very cold beer, excellent snacks).

If I ever get down there, you bet your sweet bippy I'm going to have a few cervezas at el Chan Del Agua.

Baja California

The Mexican state of Baja California is half of a 1,100-mile desert that stretches from California to the 775-mile-long state of Baja California Sur (south) that completes the peninsula. Baja California, and Sur, are only as wide as seventy-five miles, but together they're nearly as long as the American state of California. It separates the Sea of Cortez (also known as the Gulf of California) from the Mexican mainland. As for its economy, Baja California produces oranges, limes, and alfalfa, and, well, there's the city of Tijuana. I'll let your mind wander. The Sea of Cortez is a haven for whales, like the world's largest mammal, the blue whale; finback whales; and Pacific gray whales. And, there are also birds. Big, weird birds.

El Hombre Pájaro

A rural highway snakes its way through the Sierra Juarez Mountains from Mexicali to Tecate, the town of La Rumorosa (population around eighteen hundred) in the middle. La Rumorosa lies so close to the United States you could probably throw litter from a car and it would hit the state of California (not really. This is artistic license).

Four teachers saw something as they returned home to La Rumorosa on this highway in 2020, something they couldn't explain.

As they cruised on the desert highway, a creature as big as a man—with wings—soared over their car. The driver pulled over, and they watched the beast as it circled above the vehicle before veering off and disappearing into the distance. It was feathered and possessed wicked-looking claws. Its face was flat, with no beak, and the feathers surrounding the neck were of a lighter color, like a vulture, or a condor. It was el Hombre Pájaro, the Bird Man.

When they returned home to La Rumorosa, they told everyone what they'd seen, and it got on television. The teachers then discovered they weren't the only ones.

A truck driver saw the same birdlike creature near the town of San Luis Rio Colorado. Although he didn't just see it, he hit it with his truck. When he stopped to see what he'd struck, the beast stood in the road, glaring at him. It stretched its wings wide and beat them threateningly before it lifted itself into the air and vanished into the sky.

Baja California Sur

This state, as its name implies, is south of Baja California. It is, however, more of a party state. With resorts, dive sites, deep-sea fishing, and wine (lots and lots of wine), Baja California Sur is a haven for tourists. However, as far as monsters go, whereas Baja California at least has the Bird with a Human Face, Baja California Sur has...the Boy with Horrible Teeth?

The Boy with Horrible Teeth

The boy begged for money in a poor neighborhood in the capital city of La Paz. When a man approached the boy and handed him a coin, the child smiled with huge teeth, jagged like broken tombstones sticking out from rotting ground. It was the smile of a monster. The man screamed and ran home.

His story went viral at a time when that term didn't exist, and soon people saw the Boy with Horrible Teeth all over town.

Even law enforcement reported seeing the begging boy, who looked as if he were a normal child before baring bright-red gums and hideous teeth. The later encounters also involved a deep, demonic laugh. People who died soon after seeing the Boy with Horrible Teeth were considered victims of the supernatural.

What was the boy? Some suggested a monster, others a regular boy possessed by a demon. Whatever the boy was, its reign of terror only lasted a short time.

Hotel California

Warm smell of colitas…

Wait, wait, wait. *Colitas* means "pot." My word! What is marijuana doing in a rock song?

The American rock band The Eagles released the album *Hotel California* in 1976, and, although band members Glenn Frey and Don Henley claim to have written the song *Hotel California* about a Beverly Hills hotel where they both stayed, legend has it the hotel in question is the Hotel California opened in 1948 in the town of Todos Santos by a Chinese immigrant named Mr. Wong. Wong eventually changed his name to Don Antonio Tabasco, to better fit in, but the community referred to him as el Chino (the Chinese Man).

Mr. Wong also bought the first ice machine in the area, advertising the coldest beer in town, opened a general store, and the town's first gas station.

Party on, Señor Wong.

Is this building the inspiration for The Eagles's song? At this point, who really cares. What we should all care about is Mercedes, the woman who terrorizes the hotel.

Mercedes is a beautiful phantom who wanders the hotel with a bottle of wine, attempting to seduce any man she finds. If Mercedes succeeds with the seduction, her victim is doomed to death.

Mercedes has stalked the Hotel California since the 1950s, and her legend continues to this day.

Campeche

Campeche is a state in southern Mexico on the Yucatán Peninsula with 325 miles of coastline on the Gulf of Mexico, and shares land borders with Belize, Guatemala, and the Mexican states of Yucatán, Quintana Roo, and Tabasco. Maya ruins dot this state, along with ruins from the colonial era, that are UNESCO World Heritage sites. The state is flat-ish, although there are a number of high hills. Rainforests and stretches of grassland dominate Campeche, along with mangrove wetlands. In some of these areas may lurk Mexico's Bigfoot.

Quinametzin

When humans made their way into what would become Mexico (some anthropologists say thirty to forty thousand years ago, some say thirty thousand. *Tomato-Tom-ah-to*), they quickly discovered they weren't alone. Hair-covered giants occupied the lands.

These giants, known as Quinametzin, angry at the influx of these loud pesky creatures into their land, attacked the humans wherever they found them. Due to their greater numbers, humans eventually defeated the Quinametzin, driving them back into the wilderness.

Strange, but from the Americas to Australia, this story plays out across the world.

The Quinametzin are described much like a Bigfoot. The beast is between six and twelve feet tall, covered in hair, incredibly strong, and has a face much like a man, but also much like an ape. The big

QUINAMETZIN

difference between the northern Bigfoot and its southern cousin? Quinametzin have no thumbs.

The Spanish conquistadores saw these giants, as did foreigners in Mexico searching for gold. In the mid-1700s, a German missionary to the area discussed wood apes in his journals.

What were these hairy giants? The Maya considered them a spiritual part of the forest. The Aztecs considered them demons. The Spanish considered them a nuisance because these beasts would steal food from their camps, which, given their size, probably means they weren't confused with raccoons.

According to biologist Ivan T. Sanderson's 1961 book *Abominable Snowmen: Legend Come to Life*, this creature is akin to a monster in nearby Belize, the Sismite, although, unlike the Sismite, the feet of the Quinametzin aren't on backward.

Huay Chivo—Part 2

Huay Chivo is many things. He is an evil sorcerer. He is an evil sorcerer-goat. He is an evil sorcerer-deer. He is an evil sorcerer-dog. A bad, bad dog. But mostly a goat because *Chivo* is in the name, and *Chivo* is Spanish for "billy goat."

As for why the wicked Huay Chivo changes into an animal, he's hungry. After midnight, the sorcerer changes himself into a half-animal, its eyes glowing and red as brake lights, and feeds off livestock. It can appear as tall as six feet, covered in hair, with the head of whatever animal it chooses—because to change into an animal, it must leave its human head at home—then the Huay Chivo feasts on chickens, goats, and cattle.

To become a Huay Chivo, the sorcerer conducts a black magic ceremony with candles and a young farm animal it sacrifices.

Although people have reported seeing the Huay Chivo, they try not to look into those blazing eyes, for if they do, they become sick with fever.

The legend, although given a Spanish twist, is based on a Maya legend where a man fell in love with a woman of a higher class who didn't love him. Furious, he ventured into the jungle demanding to talk to a Devil. When one appeared, he traded his soul to be close to the woman. The Devil turned him into a goat.

Whereas dead farm animals had been blamed on Huay Chivo, in recent decades, the Chupacabra gets all the press.

Chiapas

Chiapas is the southernmost Mexican state, sharing borders with the states of Oaxaca, Tabasco, and Veracruz, the Pacific Ocean, and Guatemala. Its highlands are covered in pine forests, and coffee farms (with more than 400 million coffee plants). There are thick tropical jungles, Pacific beaches, Maya sites, and Spanish colonial architecture. Chiapas also produces more than 4,500 metric tons of cacao annually, so all you chocolate lovers out there, blessed be Chiapas. The state is home to diverse animals, such as spider monkeys, tropical birds (like the macaw), jaguars, and way too many reptiles. This region of 28,528 square miles (about the size of the American state of Maine) has alligators, caimans, and crocodiles—yeah, there are too many reptiles. Then Chiapas has the nerve to be home to vampire plants.

Vampire Plant

Count Byron de Prorok (although there's some debate on that "count" title) was a well-educated archeologist who explored Africa from the late 1920s to the early 1930s searching for legendary sites, such as King Solomon's Mines, Atlantis, a temple dedicated to Alexander the Great, the land of Ophir, and the tomb of Tin Hinan, the fourth-century BCE queen of the Tuareg in what is modern-day Algeria.

He actually found the last one and was labeled a grave robber.

De Prorok published four books on his travels: *Digging for Lost African Gods* (1926), *Mysterious Sahara: The Land of Gold, of Sand and of Ruin* (1929), *Dead Men Do Tell Tales* (1933), and *In Quest of Lost Worlds* (1935)—catchy titles, one and all. He also brought movie equipment and filmed his expeditions, although those films no longer exist.

In Mexico, de Prorok's legacy involves the Vampire Plant.

While in the jungles of Chiapas, de Prorok claimed to have witnessed a plant devour a bird that had landed on it by trapping the bird in its leaves and stabbing it to death with its thorns.

According to the *Encyclopedia of Cryptozoology*, de Prorok wrote: "Suddenly I saw Domingo, the leader of the guides, standing before an enormous plant and making gestures for me to go to him. I wondered what could be the matter. I soon saw; the plant had just captured a bird! The poor creature had alighted on one of the leaves, which had promptly closed, its thorns penetrating the body of the little victim, which endeavored vainly to escape, screaming mean-while in agony and terror. 'Plant vampire!' explained Domingo, a cruel smile spreading over his face. Involuntarily I shuddered; the forest was casting its evil spell upon me."

With more than 750 recognized carnivorous plants on Earth, like the Venus flytrap, the sundew, cobra lily, various types of pitcher plant, and the best name, the bladderwort, what's one more?

The Alux

You may not see them coming. In fact, you probably won't, unless they want you to.

An Alux is the little person of the Yucatán Peninsula that protects, pranks, or murders humans depending on the human, and the Alux's mood. Similar creatures appear in mythology across the world: brownies, Duende, Ebu Gogo, elves, goblins, leprechauns, Menehune, Pixies, Pukwudgies, Redcaps, sprites, Tomtars. Call them what you will, they are all around two to three feet tall, can turn invisible, are mischievous and deadly. They dress like the people in the areas they inhabit.

These little people can be found in caves, forests, limestone sinkholes, and around large stones (much like elves in Icelandic folklore). If farmers build a small house for the Alux, some may move in and provide rain for crops and protect the land when the farmer is asleep. However, if the Alux stays too long in the house (about seven years), it will grow bitter and stop protecting the fields, preferring to cause chaos instead. Maya legends say sealing the house traps them inside and keeps them from mischief.

Today, on the Mexican states of the Yucatán Peninsula and in the nearby countries of Belize and Guatemala, the Alux is sometimes blamed for natural disasters, construction collapses, or plagues, usually in response to an affront to the Alux, such as moving an Alux stone to build a road or bridge.

In 2023, Mexican President Andrés Manuel López Obrador posted a photograph of what he claimed was an Alux on his social media accounts.

"Was taken three days ago by an engineer, it appears to be an aluxe," he posted along with the photograph. "Everything is mystical."

Dzulúm—Part 1

The white jaguar is another creature of Chiapas that came from the world of the Maya. The creature is always a huge male with the long mane of a horse streaming from its head. When it encounters a local woman, the spell of the Dzulúm attracts the woman to it, and the woman follows and vanishes, sometimes forever, sometimes not. Those women who want to live make a deal with the Dzulúm, trading it their soul in exchange for the powers of witchcraft.

In the wilderness, other cats are said to offer the Dzulúm sacrifices because they are terrified of it.

In the *Bulletin of Hispanic Studies*, volume 88, number 6, an article by Chris Harris entitled "The Myth of the Dzulúm and Patriarchal Masculinity in Rosario Castellanos's Balún-Canán" posits the Dzulúm is a reflection of the dangers of the patriarchy. The actions of the Dzulúm, the article claims, reflect masculinity's destructive impact on society for both women and men.

No Dzulúm could be contacted for comment.

The Werewolf of Coita

When residents of Ocozocuautla de Espinoza, Chiapas, began to hear howling in 2019, they didn't expect the kind of creature it came from. Then, following an unsubstantiated account of a woman attacked in the dark, people spread the word—there was a werewolf in town.

In April 2020, in Ocozocuautla de Espinoza (known locally as Coita), the howls were followed by gunshots, then claims from residents of a hairy, fowl-smelling, two-legged beast, at least six feet tall, with the ability to leap over high fences.

On April 11, 2020, @saulzenteno posted on Twitter (at that time it was still Twitter):

"I don't know if you knew but in Mexico there is an ENTIRE town that hasn't slept for two nights because there is a werewolf. Coita, Chiapas. #COVID19 will bring us harm but the werewolf in Coita is another level. *#MéxicoMágico* #Real #NoFake #cuarentena."

Then police got involved, and they saw it too. While pursuing the creature, it leaped a six-foot-tall fence, then onto a rooftop, then it escaped.

Some people reported seeing two or three beasts, but then the sightings stopped.

What was the Werewolf of Coita?

Speculation revolved around natural creatures being forced out of their natural habitat by human encroachment. Other thoughts were more spiritual, enough so a local priest advised lighting candles to ward off evil. However, that area of the country has a long history of werewolf-like beings called Nagual. This is a witch that can transform itself into its spirit animal, which is, of course, sometimes a wolf.

Whatever the creature was, it left footprints. Big, canine footprints.

Cax-Vinic

Wanderers in the forest first hear the Cax-Vinic's warning—a deep, long, terrifying howl. This wild man wanders the wilderness at night in the forests of Chiapas; its glowing eyes may be the next warning a person sees, before it emerges from the trees, an apelike human covered in dark hair.

Biologist Ivan T. Sanderson documented the beast in his book *The Abominable Snowman: Legend Comes to Life*, writing the Cax-Vinic are "large, hairy men who live in an immense, unexplored canyon in the Sierra Madre Occidental"—and the locals are terrified by them.

The Cax-Vinic (also known as Cangodrilo, Fantasma Humano, and Hombre Oso) are often reported in the higher elevations, local residents claiming to find bare humanlike footprints in the snow up to 6,500 feet, according to a 1937 article by W. C. Slater in the *Times of London*, "The 'Abominable Snowmen': Footprints in Mexico." The villagers saying the prints were from "the men of the snows."

Chihuahua

El Estado Grande—the Great State—is called such because it is the largest Mexican state at 95,543 square miles, about the size of Oregon, and is one of six states that border the United States with 583 miles of border with New Mexico and Texas. Although Chihuahua is known for the Chihuahuan Desert (the largest desert in Mexico), it has more forests than any other Mexican state including the largest conifer forest. The Sierra Madre Occidental Mountain Range (we don't need no stinkin' badges) runs through Chihuahua, along with las Barrancas del Cobre (the Copper Canyons), which, although not the deepest canyon in the world, it is the largest canyon system on Earth. Man, oh, man. Chihuahua has a lot of places for cryptids to hide—even big ones.

The Copper Canyon Monster

One of the largest tribes of Indigenous people in North America, the Rarámuri (*Tarahumara* in Spanish) is composed of around one hundred thousand people who live in Chihuahua's Sierra Madre canyons, and the Sierra Madre Occidental Highlands.

According to the website *Mexico Unexplained*, the Rarámuri also live with a beast—an insatiable beast.

This humanoid monster is unusually tall with bones protruding from skin pulled tight over its thin frame. Its fingers are long, and sharp, its teeth are long...and...sharp. It has antlers, and hunts and eats people, and is....

Hmm. This really, really sounds like a Wendigo.

The Copper Canyon Monster can climb trees and cliff faces with ease; it can run with the fastest Rarámuri. The Rarámuri are legendary for their long-distance running, with claims that people of that tribe can run for more than 200 miles without slowing, and one person purportedly ran 435 miles in two days. Their feats of feet are done in simple flat sandals, or without any footwear at all. Their story is interesting, but this is about the monster. Google them.

Slinking through the canyon system, no fur to warm it in the cool nights, it is protected by its impenetrable skin and fierce antlers.

When it comes to humans, it can either eat them or possess them. A report from the 1800s tells of a Rarámuri man whose mind was reportedly taken by the monster. He killed his family. Much like the legends of the Wendigo, the man claimed his family starved to death, and eating them was the only way he could survive.

Like the case in Fort Kent Alberta, Canada, authorities claimed this man used the Copper Canyon Monster as an excuse for murder.

Giants, Giants, and More Giants

The Rarámuri didn't only have the Copper Canyon Monster in their belief system, they had the Ganoko giants.

Much like other parts of Mexico, when the Rarámuri settled Chihuahua, the Ganoko giants already dwelled there. However, unlike other parts of the country where the humans and giants warred, the Rarámuri and Ganoko were peaceful, even farming together, according to an article in the *Chihuahua Post.*

However, when the Rarámuri gave the Ganoko food and fermented drink, the giants ran amok in a rampage of destruction and cannibalism. To combat them, the Rarámuri poisoned the giants.

From the May 1998 *Smithsonian Magazine*:

"In the 1890s, Carl Lumholtz was told a legend about a race of giants that had occupied the canyon country when the Tarahumara (Rarámuri) arrived. The giants ate the Tarahumara children and ravished the women. At last, the people exterminated the giants by tricking them into eating a mixture of corn and a poisonous extract from the chilicote tree."

Did these giants exist? Rarámuri cave paintings depicting the Ganoko were discovered in Guerrero, Chihuahua, in the 1980s, so, yeah. Maybe.

The Starchild

American paranormal researcher Lloyd Pye died on December 9, 2013, but while he was alive, he thought he was on to something big.

A girl wandered into an old mine tunnel in Copper Canyon in the 1930s and discovered a human skeleton. Beside it was a child's skeleton—the skull was strange, large, bulbous. The skull was passed down for years, until Pye acquired it from Ray and Melanie Young of El Paso, Texas.

The skull fascinated Pye.

The brain cavity is larger than that of the average human adult's at 1,600 cubic centimeters. The average for males being 1,481.6 cubic centimeters, and the average for females at 1,375.4. The range of a "normal" adult's brain cavity, regardless of sex, is between 1,203 and 1,678 cubic centimeters.

STARCHILD

So, yeah. Its brain capacity is big.

There are other abnormalities. Its eye sockets are shallow, the back of the skull is naturally flat, there are no sinuses. The bones are light, about half the weight of human bones, and half as thick, although stronger than human bones.

DNA testing showed the mother to be human, but daddy? According to Pye, Dear Ol' Dad's story is a mystery, although analysis from Lakehead University in Thunder Bay, Ontario, Canada, determined the Starchild to be 100 percent human.

Coahuila

This state in northern Mexico is a yin-yang of geography, with the Chihuahua Desert in the west, and temperate highlands in the east. Coahuila borders Texas for 318 miles along the Rio Grande (Río Bravo del Norte), which is understandable, because the Mexican state Coahuila y Tejas included Texas from 1824 until Texas declared its independence from Mexico in 1835. However, despite Coahuila's long and storied history, I'm going to focus on food. In 1943, Ignacio Nacho Anaya García, in the border town Piedras Negrasis, served a plate of tortilla chips covered in melted cheese in the restaurant the Victory Club. The guy created nachos. I should stop writing this section immediately. How could I top that?

Mexico's Mothman

Mothman. The name conjures thoughts of Point Pleasant, West Virginia, USA, in 1966, when two young couples were making out in a car in the TNT Area outside town and saw a humanlike beast. It was more than six feet tall, with no neck, blazing red eyes, and wings with a span of at least twelve feet.

More people eventually saw the flying creature a newspaperman dubbed the Mothman, although it didn't look like a moth; it looked more like a man-bird. The sightings were coupled with Men in Black encounters, and a claim that the creature portended Point Pleasant's Silver Bridge collapse, killing forty-six people on December 15, 1967. The creature was immortalized in paranormal journalist John Keel's 1975 book, *The Mothman Prophecies.*

Although the Mothman seemed to disappear after the Point Pleasant sightings, people have reported seeing it, or a similar creature, before other disasters, such as the 1986 Chernobyl, Ukraine, nuclear power plant meltdown, the 2007 Interstate 35 West bridge collapse in Minneapolis, in which thirteen people died, and a spate of sightings in Chicago starting in 2011.

There weren't any real disasters in Chicago that year, but the Cubs record was pretty bad.

What does this have to do with Coahuila? In April 2024, in the Castaños municipality (equivalent to an American county/parish/commonwealth), a family claimed to have photographed a Mothman-like creature flying over the desert.

The figure was winged, humanoid, and dark against the clear evening sky. According to the *Tamaulipas Post*, the man-size creature sprang from the shadows and terrified all who saw it before it disappeared into the distance.

Did it foretell a disaster in Castaños? Not as such.

According to Tim Binnall, news editor at *Coast to Coast AM*, the photograph of the Castaños Mothman that went viral on social media in Mexico was really a picture claimed to have been taken in Point Pleasant in 2016.

The story itself grew wings, but don't trust social media for accuracy, folks.

Monclova Monster

Over a five-day period in 2010, numerous residents of the east-central Coahuila town of Monclova (population 260,000) saw a creature they couldn't explain.

According to researcher Lon Strickler at his website *Phantoms and Monsters*, on April 5, two thirteen-year-old boys on Calle Londres (London Street) were terrified when they saw a four-legged, hairless, gray-skinned creature with the features of a man. When the children rushed home, one of the mothers called the police.

Officers went to Calle Londres in the Chinameca district, but the creature was gone.

The next day at around 2:00 p.m., swimmers saw the beast underwater in the Rio Monclova. Although officers were again summoned to the site of the encounter, all they could do was interview witnesses and tell swimmers to get out of the water. The creature had, again, escaped.

That same night, two teenagers reported seeing the creature as well.

The last sighting came three days later. A man from Ejido Curva de Juan Sánchez, in the southern part of Monclova, told police the gray humanoid creature killed ten of his goats during the past few months.

What is this creature? Sounds like the Dover Demon or the Loveland Frog, both covered in *Chasing American Monsters*.

Colima

Colima is small, but its land is diverse, as is its wildlife. From the humid coastal plains to the foothills of the Sierra Madre Occidental Mountain Range, Colima is home to ocelots, deer, wild boar, and enough snakes to give you the willies. With its one hundred miles of Pacific coastline, it's also a haven for tourism, welcoming around six million visitors per year. Colima is also known for its two volcanos, Nevado de Colima and Volcán de Colima. You know the state motto? "Come for the high standard of living and low unemployment, stay for Volcán de Colima, one of North America's most active volcanos!" (Not the real motto. That would be, "The spirit of the arm is force on Earth.") Oh, and Colima also has a flying horse. Stay for that, too.

The Flying Horse

Smoke rolled from Volcán de Colima in 2015 when twenty-four-year-old Loyola Quintanilla Rosas saw something strange—a black shape resembling a horse hanging in the sky near the volcano, according to an article in the *UK Daily Mail.*

"When I first saw it, I thought it looked like a horse. It had a very large body and seemed to be an animal," Rosas told the *Daily Mail.* "It had a thin top, a bulging middle and two extremities at the bottom, but the legs weren't moving so it clearly wasn't a modern-day Pegasus."

As fast as it appeared in the turbulent sky, it vanished.

Paranormal researchers and government officials both took turns at explaining the flying horse.

For paranormal researchers, the answer to the horse could be anything from an undiscovered animal to a UFO. For Mexico's National Civil Protection System, "It is most likely a drone of some sort. We are looking into it."

Regardless of what the black shape was, Rosas captured it in a photograph; the debate can continue.

Chupacabra

This four-foot-tall, red-eyed, spike-backed bloodsucker has terrorized Latin America since Madelyne Tolentino saw this creature in August 1995 in Canóvanas, Puerto Rico (see the full report in the chapter Puerto Rico). It didn't take long for the entity called the Goat Sucker to make its way across the Caribbean to Mexico and the Southern United States.

Attacks were all similar. Farmers would discover that, during the night, small livestock—goats, turkeys, chickens, and rabbits—had died; they were drained of blood. Then attacks escalated to larger animals, such as cattle and horses.

Attacks of a Chupacabra reached as far as the Pacific Coast city of Manzanillo, Colima.

Reports of dead goats became common around Manzanillo. Puncture marks in the neck and skull killed the animal before it was drained of blood. No predator would touch the animal after the Chupacabra fed, and the owner would discover their livestock in the morning, wondering what had happened.

In most claims of a farmer or rancher capturing or shooting a Chupacabra, the creature is usually a raccoon, dog, or coyote with mange.

There is a Colima-related explanation cryptozoology skeptics have used to explain Chupacabra sightings—the Xoloitzcuintle, the Colima dog.

For three thousand years, the Xoloitzcuintle were raised, loved as pets, and harvested for meat by the mostly vegetarian people who lived in the Colima region. Xoloitzcuintle were loved for their companionship and loyalty, and clay figures of the dogs were often placed in graves to give the deceased protection, and (cough, cough) lunch, for their trip to the afterlife.

The Xoloitzcuintle does look like a dog with mange, although I doubt the Chupacabra-blamed livestock deaths could ever be blamed on a Xoloitzcuintle. From the one I saw, she looked like a good girl. Oh, yes she did. Good girl.

Durango

With its desert, mountains, and 12,108,164 acres of forestland, the state of Durango looks like a Western. That's a big reason why John Wayne movies, like *The Sons of Katie Elder* (1965), *The War Wagon* (1967), *Chisum* (1970), and *Big Jake* (1971), were all filmed there. Durango is Mexico's fourth-largest state at 47,631.1 square miles, but twenty-fifth in population with 1,832,650. The capital city, Durango de Victoria, dates back to Spanish colonial days, and is considered a resort city due to its abundance of hot springs. Mexico named the city a national monument.

The Chicken Killer of Lerdo

The small town of San Luis del Alto (population fewer than 1,000) is in the far north part of Durango, just south of the much—much—larger city Gómez Palacio (population around 342,286).

In August 2023, farmers and ranchers in the high desert between San Luis del Alto and Cañon de Fernández State Park discovered chickens, sheep, and goats on their properties were dead. Uneaten, but dead. What troubled the farmers, more than the deaths, was the evidence of the deaths. There was none. No blood, no tracks, no damage to fences or cages.

According to the website Mexico Unexplained, different animals had wounds in different spots. Small puncture marks between the front legs of goats and on the necks of sheep went into the animal up to a foot; the marks on chickens punched through their chests. All the blood was drained from each animal, and every wound appeared cauterized. Some animals had marks on their bodies that looked caused by the suckers from an octopus's tentacles.

No animal in the region killed like that.

What was it? Some farmers and ranchers reported seeing lights—"bad lights"—in the nearby hills. Others reported seeing el Hombre Pájaro, the Bird Man, as has been reported from the Sierra Juarez Mountains in Baja California. Another reported seeing a bipedal cat man walking across the ranch land on the night of an attack.

Although the attacks are similar to those of the Chupacabra, there's enough evidence the people of the area believe the deaths were caused by something else, something they don't yet have a name for.

Tall Aliens of la Zona del Silencio

La Zona del Silencio—the Mapimi Zone of Silence—has a strange history, and present, and probably future.

Bolsón de Mapimí is a fifty-thousand-square-mile spot of desert where what water it gets drains toward its center into lakes and swamps as opposed to outward toward the sea. In that spot of desert is the Zone of Silence, a patch famous enough to be marked on Google Maps. That makes it famous, right?

Let's take a step back to the 1930s.

A pilot, flying over Bolsón de Mapimí, an area already known as a hot spot for meteorites, realized his radio had died. It didn't begin working again until he left the area. Locals and visitors discovered this happens all the time. When someone enters a zone in Bolsón de Mapimí, radios stop working, and compasses go haywire. Today, mobile phones are useless in la Zona del Silencio.

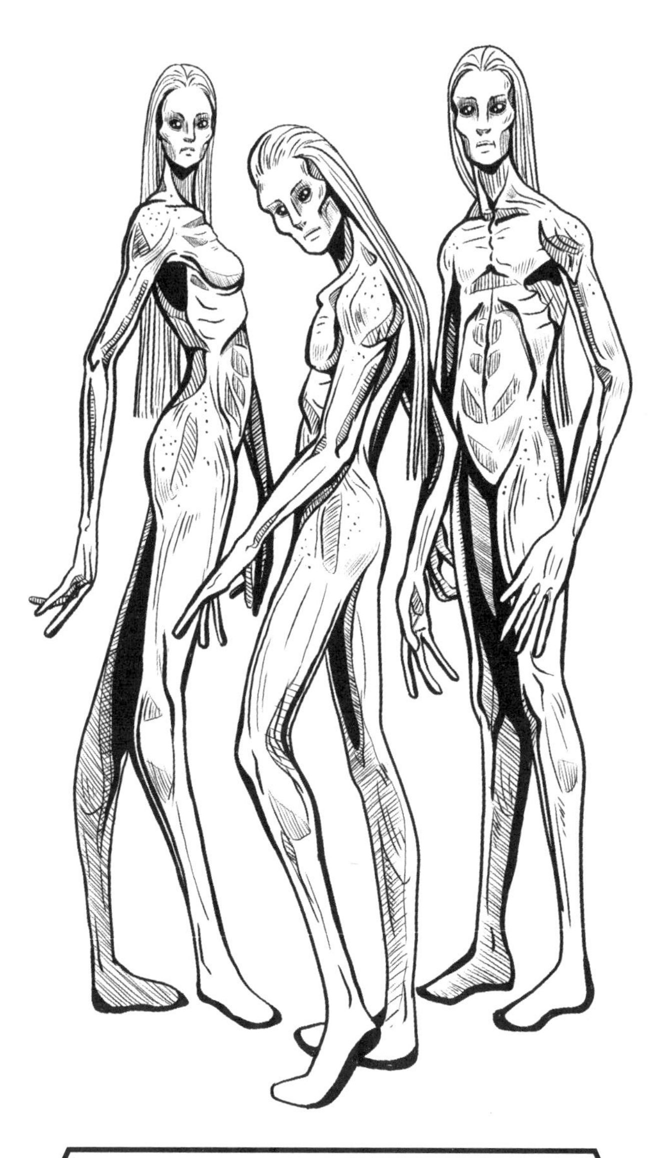

TALL ALIENS OF
LA ZONA DEL SILENCIO

On, July 11, 1970, NASA launched an Athena rocket from the Green River Launch Complex, at Green River, Utah, with equipment to study the upper atmosphere. The rocket flew off course and crashed into, you guessed it, la Zona del Silencio. The US government sent a team to recover the remains, including, in some stories, former Nazi scientist and leader of the US Space Program Wernher von Braun. The team recovered the materials and returned them to the United States.

UFOs are reported from la Zona del Silencio, as are strange lights and fireballs flying through the sky or rolling across the desert. Some suspect the electromagnetic interference is caused by materials in meteorites buried in the sand. Others believe it's because of the tall aliens.

The stories are similar. Visitors venture into the Zone of Silence and get lost (does your GPS work? Nope), or stuck, and unusually tall blonde people appear and help the misled find their way, or assist the stuck out of their jam. All they ever ask for is water, if anything. And if asked where they came from, they only point toward the sky. Then, the tall blonde visitors vanish. No goodbye, no vehicle they could have left in, no footprints. They're simply gone.

La Lechuza and the Witch of Durango

La Lechuza, the Owl, is a deep-seated Mexican legend, but this owl isn't an owl; not really. It's a witch, a bruja, in the form of a giant white owl with an old woman's face—and la Lechuza is out for revenge.

This anthropomorphic bruja hunts in the night by luring people to them with a baby's cry, or it whistles at its victims. Then it eats them. Seeing one, or hearing its whistle, portends death. It's hard to kill, unless, of course, it's trapped first.

In 2014, villagers in rural Durango captured a white owl, believing that owl was a brujo, and set it on fire.

According to an article in the *UK Daily Mail*, villagers interrogated the owl—its legs tied with rope—the villagers demanding it turn back into a woman. The injured owl shrieked at the women who questioned it. The women interpreted "the bird's cries as proof that it is really a witch and does not like the fact that its true identity has been discovered."

They read Bible passages to the owl, then promised to release it if it repented, but the owl remained just an owl.

The owl's transgressions included looking at a villager through a window and placing a curse on another.

It never, of course, turned into a woman, nor did it apologize.

The *Daily Mail* claimed the owl survived the attack, and the villagers received a worldwide negative response, but this doesn't mean anyone listened. In 2019, two Durango teenagers apparently burned a white owl they suspected of being la Lechuza.

Hey, folks. Sometimes an owl is just an owl. And by sometimes, it's always. Well, unless it's a little gray alien masquerading as an owl. That totally happens.

Guanajuato

The state of Guanajuato is known for silver mining. No, no. Independence. It's known for the first battle for Mexico's independence from Spain. Wait, wait, wait. Mummies. Guanajuato is known for mummies. Bodies unearthed in the capital city of Guanajuato between 1870 and 1958 have been naturally mummified, and people, curious, started paying money to see them. This led to the city opening el Museo de las Momias (the Museum of the Mummies) in 1969. That's still not right. This central Mexican state is known for its more than one thousand volcanos as part of the Trans-Mexican Volcanic Belt. Wow. I guess it's also known for giant reptiles.

The Snake Princess of Bufa

On Holy Thursdays—the fifth day of the Christian Holy Week—a Native princess can sometimes be seen wandering the hills at the southeast edge of the city of Guanajuato.

Someone will probably see her again. They have for the past four hundred years.

Legend has it, in the 1600s, a sorcerer cast a curse on an Indigenous princess, dooming her to a half-life/half-phantom existence unless she finds a handsome young man brave enough to break the spell. To do so, the man must escort the princess to the altar of what is now the Basilica of Guanajuato, a four-hundred-year-old Catholic church in the nearby city.

Praying at that altar is the only way the woman can once again become fully human, and, if she ever does, her savior becomes a rich man.

However, there's a specific way the man must deliver the princess.

Once he answers her call, he has to carry her, ignoring all stares from passersby. Disembodied voices will assault the man once he enters the city with the cursed princess; he must ignore them, never looking around to see where the voices are coming from, never straying his eyes from the church. If he panics, and tries to return the princess to the hills, he'll wish he'd had more bravery because the princess transforms into an enormous snake and devours her failed savior.

No man has succeeded in reaching the church with her. Yet.

Valle de Santiago Lake Monster, UFOs, and Stuff

A lot goes on in Valle de Santiago (Santiago Valley). UFO sightings, local farmers growing enormous vegetables, and, of course, reports of a lake monster.

At least, there used to be reports of a lake monster back when there was a lake.

Rincón de Parangueo (Parangueo Corner) is a small town (population 3,015) in the valley, near the city of Valle de Santiago. A quarter mile from Rincón de Parangueo is the Crater el Rincón de Parangueo (Parangueo Corner Crater), once home to the Valle de Santiago Lake Monster.

Until the late 1980s, this volcanic crater was filled with water, but then, local farmers pumping groundwater drained it. However, that doesn't mean the crater's not interesting. The white sandy landscape is dotted with green, and there are vendors looking to make money from gaping tourists by selling food and postcards. The crater is dry and beautiful. Why wouldn't it be; it's claimed to be one of the Seven Luminarias, seven local volcanic craters that, from the air, are in the form of the Big Dipper.

Did I mention visitors reach the crater through a fifteen-hundred-foot man-made tunnel? That's not creepy. Not at all.

As the stories go, the water the crater used to hold would occasionally turn red, as locals tell, to predict natural disasters, such as the 1985 earthquake that rocked Mexico City.

As for the monster, for decades, visitors claimed to have seen a Loch Ness–like monster swimming in the lake, according to the *Guadalajara Reporter.* Although reports since the 1980s have—cough, cough—dried up, plesiosaur fossils have been discovered in Valle de Santiago, so, who knows.

Guerrero

Although Guerrero is a producer of coffee, cacao, tobacco, and bananas, gold, silver, and copper, to Americans, this Pacific state is usually associated with one word: *Acapulco.* This resort city is located on Acapulco Bay, the Sierra Madre del Sur Mountains looking over the city's shoulder. It is home to hiking, golf, dining, and watching amazingly brave divers throw themselves into the sea off one of two ledges of the la Quebrada Cliff: one at forty feet, the other at eighty. Yikes. But the state is more than just tourism. There are mountains, canyons, forests, and plains, filled with snakes, iguanas, eagles, and jaguars. Guerrero also has a rich Aztec history. There's also the Chaneques.

Chaneques

The Chaneques are little people who've been around since the Aztecs ruled Guerrero. Like the European sprite, these tiny beings are part of nature, acting as the guardians of rivers and forests. Although their appearance differs per region, they are most often depicted as children with the faces of old men and women.

As for their behavior, stop me if you've heard this before.

Chaneques are known to make people lost in the wilderness—for days (so do elves, sprites, Pukwudgies, etc. Yep, sounds familiar). Sometimes they take these people to their realm (so do elves, sprites, Pukwudgies, etc.), and when the people return home, they remember little of their adventure (like when they go with elves, sprites, yada yada). Sometimes, the kidnapping is simply to help the Chaneques procreate (sorry. I hate repeating myself).

Sometimes, the Chaneques do favors for people instead of misleading and kidnapping them. That is, if the people they encounter respect nature (like the...oh, forget about it). Some people (the smart ones) give Chaneques a tribute to ward off mischief, and to ask the Chaneques to protect their homes (like people in other cultures do for those other little guys). You know, I'm beginning to seriously believe fairies are real.

Wait. Hold the phone. There is *one* big difference between all these little people legends—instead of through boulders or fairy circles, the way into the Chaneques realm is inside a dried kapok tree.

Okay. Now they seem totally different from all the others.

The Blob of Bonfil Beach

It didn't look appetizing, but, then again, I'm not a fan of seafood.

In 2016, tourists walking along Bonfil Beach in Acapulco stumbled upon something...nauseating. The rotting corpse of a thirteen-foot-long sea beast had washed up on a spot where people take their children to build sandcastles while they drink too much, and might have sex on that same stretch of beach later. I mean, that monster committed a diverse range of beach party fouls.

Tourists, of course, did what tourists do—took photos and put them on Facebook, or, what was around in 2016? Google+, maybe?

Local authorities didn't believe the creature had been dead for long, according to a story on *Yahoo! News*, although if it had recently died, it decayed fast, but it decayed strangely.

BLOB OF BONFIL BEACH

"We have no idea what type of animal this is," Rosa Camacho, of the Civil Guard and Fire Brigade, told *Yahoo! News*. "But I do know that it does not smell bad or have a fetid aroma."

Hmm.

Although the photo was shared around the world, no one has been able to tell what kind of creature it was. Whale? Squid? Plesiosaur? Cthulhu?

Wait.

In March of that year, the curator emeritus of marine mammals at the National Museum of Natural History in Washington, DC, said he thinks the blob is the top half of the head of a sperm whale.

You, dear reader, have to ask yourself this: Do you trust the judgment of an expert who based his opinion from a photo on Facebook. I mean, Facebook. Come on.

Hidalgo

Hidalgo is one of the smaller Mexican states. It's just north of Mexico City and was once home to the Toltecs. The culture of the state is varied, from the traditional Indigenous cultures, to the Spanish influence, to that of Italian immigrants, Jewish immigrants, and famed miners from Cornwall, England, who came for silver. Hidalgo is mostly mountainous, although there are plains, hot water springs, and the ruins of Tula, the capital city of the Toltec Empire. With mountains, forests, and plains, Hidalgo is a perfect environment for opossums, whitetail deer, porcupines, wild turkeys, jaguars—and mysteries.

Charro Negro

A lot of monsters only appear in darkness, and Charro Negro is no exception. He is a reason for people to be at home, and for mothers to hug their children tight.

This entity's story didn't begin as a nightmare. He was a man, born to a poor, loving family, but hoped for better. As he grew, he saved what little money he earned to buy flashy clothing and fancy hats. Being poor, however, was not what this boy wanted. He wanted riches. In short, he was high maintenance.

After his parents died, he did what any honest, self-respecting person would do to better their life. No, he didn't get a job; he summoned the Devil.

The Devil appeared to the boy, who was now a young man, and the man begged this demon for money, and the Devil gave him money—in return for his soul (come you. You saw where this was going).

The man started spending his ill-earned cash, and didn't stop for a long, long time, because his supply of money never ran out. He spent it on houses, on wine, on women, and, of course, on fine clothing and hats. Nothing was too good for him, and nothing was too much—until it was.

The Charro grew old, and realized he was only the shell of a person, a mask, an open wallet for "friends" who would leave if he were suddenly broke. Worse, it would soon be time to pay the Devil. Frightened of an eternity in Hell, he did all he could to avoid the Devil. He hid, placed Christian emblems all over his property, and prayed in a specially built chapel. But the truth weighed on him. The Devil would eventually come to his house and take his soul.

So, dressed in his best black suit, he mounted his favorite black horse and fled.

He couldn't flee fast enough, or far enough. The Devil found him and sucked the soul from the Charro, leaving only a living skeleton inside the man's best suit, still astride his best horse. Then the Devil gave him a job to collect the debts the living owed the demon. If the Charro Negro ever encounters a man that is as greedy in life as him, that man will take his place.

This apparently hasn't happened because the legend endures.

If you're ever in Hidalgo and are caught in the countryside at night by a black horse ridden by a red-eyed skeleton dressed in black, you're probably doomed to spend an eternity in Hell.

On the lighter side, Charro Negro is also a cozy restaurant in Pachuca, Hidalgo, with a five-star Google review, located between Home Depot and Walmart.

Home Depot and Walmart. I am not making that up.

Orb Witches

Will-o'-the-wisps occupy a place in mythologies around the world. They're balls of light that lead unsuspecting wanderers on a trip through the forest, getting them lost, sometimes forever. In real-life science land, they could be caused by "swamp gas." However, lights such as these don't always appear in swamps, and sometimes they seem to be sentient.

In Hidalgo, there's a long tradition of lights such as this that lure people into the wilderness, usually children, and, instead of simply getting them lost, these vampiric entities target children to suck their blood.

These orbs are actually witches.

The legend starts with a farmer's wife villagers believed was a witch who cooked her husband's food with the blood of babies. Turns out she didn't. However, when the farmer lay down, pretending to be asleep, he witnessed his wife remove her leg and transform into a turkey, then fly out the window of their hovel. When he went to the window, the turkey transformed into a ball of light.

He hid the leg, and his wife couldn't return to human form. The turkey was then burned in the village center.

Since then, any unexplained fiery light in the night has been blamed on a witch-turkey-vampire thing.

Jalisco

If Mexican states were nonverbal communication, Jalisco would be the high five. If it were a playing card? The Ace of Hearts. No doubt. Tequila was invented in there, as were rodeo, mariachi music, and the sombrero. Oh, and three-time Academy Award–winning director Guillermo del Toro is from Jalisco. Couldn't you share with the other states? It also has the popular vacation spot Puerto Vallarta, beautiful Spanish colonial architecture, and Lago de Chapala, the largest freshwater lake in the country. Seriously? Two mountain ranges move through Jalisco, as do forests and plains. It also has 213 miles of Pacific coastline. I may move there. Oh, wait. Yeah. There's the dragon, and giants, and the Thorny One. Maybe not.

The Tequila Dragon

The Tequila Dragon is not a drink (although it will be by the time I finish this book. That is called research and development).

The city of Tequila, Jalisco, (population 40,697) is situated between Puerto Vallarta and Guadalajara and is surrounded by fields of blue agave (for, of course, making tequila). Visitors can ride the Jose Cuervo Express train through the valley, mountains, and agave fields, sleep in a tequila barrel at Matices Hotel de Barricas (the hotel has a bar. Was there ever any doubt?), and visit tequila distilleries...Okay, we get it, lots of tequila. Let's get to the dragon.

Parroquia Santiago Apostol is a Catholic church a ten-minute walk from the city's center (or AutoZone, whichever you prefer). Built in the eighteenth century, this chapel is, well, it, um, it has an enormous lizard.

Legend has it a dragon sleeps beneath the town of Tequila; its tail is under Volcán de Tequila (Tequila Volcano), and its head is beneath the church. There's apparently a breeze inside the church people have attributed to the breath of the sleeping dragon. As the story goes, a violent storm struck Tequila, and while the townspeople sought safety in the chapel, the priest walked out into the storm, and the storm dissipated.

Throughout the years, the dragon and the storm became one, and the priest tamed the dragon, putting it to sleep. It still sleeps. A few Tequila residents still dread the day it may wake again.

Giant Rocks

Enormous stone spheres have been discovered in various parts of the world, such as Bosnia and Herzegovina, New Zealand, and the Orkney Islands of Scotland. The most famous are in Costa Rica; however, there are also stone spheres in the Sierra de Ameca mountains near the city of Ahualulco de Mercado (population 23,630). Ideas of how these spheres formed usually involve being carved by an ancient civilization, or space aliens. Scientists come up with their own explanations, mainly because they can never let anyone have any fun.

The giant spherical rocks of Jalisco? They were made by giants. I am not taking any questions at this time.

These rocks, anywhere from two to eleven feet in diameter, were buried in volcanic ash for a ridiculous amount of time (I'm going with the giant hypothesis, so there) and were discovered in modern times when taxi driver Miguel Hernández noticed the tops of them uncovered after a violent rainstorm in the 1960s, according to *Mexico News Daily*.

Called las Piedras Bola (literally, "the Ball Stones"), the local legend is that the more than seventy-two rocks were balls giants tossed to each other in play. When the giants left, they didn't take their balls and go home. Geologists, however, suggest they're a natural formation formed from crystalized volcano ash.

Regardless of how they were formed (or who played catch with them), Jalisco's Ministry of Culture spent more than ten million pesos to turn the site into a tourist attraction. It has as a road (visitors used to have to walk four miles to see the stones) a suspension bridge and, of course, zip lines because, why not?

Ahuítzotl

People upon the water—travelers, fishermen, swimmers—don't know the Ahuítzotl lurks nearby until it strikes. This beast (named after the Aztec emperor from 1486 to 1502 CE) sports small ears and short fur; its tail is long, and black, and tipped with a human hand. The tail's hand grabs anyone infringing on the Ahuítzotl's territory and drags them beneath the water to their death.

That's not all. The Ahuítzotl rips out its victim's teeth, nails, and eyes. When the body surfaces, only priests can remove it from the water because the creature is a friend to the gods of rain. Because of the beast's relationship to the gods, the spirit of anyone good it kills goes to the paradise Tlalocan. Anyone bad is punished.

This creature is also called the Thorny One of the Water and lives in the lakes and rivers of the ancient Aztec empire. "Thorny" because when it leaves the water and shakes itself off, its fur becomes bristles.

Was—or is—the Ahuítzotl real?

People have speculated it may be a tlacuache, but, since that's really a water opossum, I gotta say…

A water opossum? Sure, the description is near the mark, and it has a prehensile tail that could be mistaken for a hand (maybe?), but, at most, it's thirteen inches long with a sixteen-inch tail. Not really capable of dragging an adult underwater.

The Itzcuintli? This hairless dog looks the part, except for the thorns and hand on the end of its tail. Others suspect an otter (who can be violent toward humans), or, most probably, something mythical. Either way, I'm not swimming in Jalisco anytime soon.

Mexico City

The capital of Mexico, Mexico City, is big. Big, big, big. It's not only the largest city in North America (with a population of 21,671,908), the metropolitan area is only behind Tokyo, Delhi, Shanghai, and São Paulo, for the title of biggest city on the planet. It includes Teotihuacan, the Aztec City of the Gods, that features the Pyramid of the Sun and the Pyramid of the Moon. In December 2004, Walmart opened a store nearly in the shadow of the pyramids, which threw the world into a tizzy, understandably. That Walmart is now closed. This city loves history. With 173 museums in Mexico City, if a person took two hours per museum, it would take them 346 hours (14.5 days straight) to properly visit each one. If they did one per day, it would take half a year. I would totally do that tour. The Basilica de Santa Maria de Guadalupe in the city is the home of Mexican Catholics. Mexico City also offers world-class architecture, world-class food, and world-class entertainment. Really? Do you need anymore? How about flying people?

The Policía and the Bruja

Police officer Leonardo Samaniego Gallegos cruised Alamo Street at about 3:15 a.m. on January 16, 2004, when a black figure descended from a tree about twenty feet off the sidewalk.

The figure was a woman, a floating woman.

She swooped down to the patrol car, her black cloak billowing behind her, her eyes black voids. The woman slammed into the patrol car's windshield and cracked the safety glass. She clawed at the cracks with long sharp talons.

Frightened by this hideous being, Gallegos slammed the transmission into reverse, and, to escape, floored the accelerator, speeding backward down Alamo Street. He managed to call for backup before crashing the car into a wall.

Now stopped, the creature hit the windshield again, clawing and gnashing its teeth at Gallegos, her empty, lidless sockets glaring at him. He closed his eyes and prayed before fainting.

When backup and an ambulance finally arrived, the being was gone, but Gallegos didn't hesitate to tell of his encounter with the *bruja* (witch).

Gallegos wasn't the only one to see this flying entity. Three police officers from Santa Catarina—six hundred miles away—saw it on January 13, but were afraid to admit it until Gallegos came forward. In 2006, residents of Monterrey also reported seeing a similar woman flying in the daylight.

Monster of Madín Dam

Despite years of pollution violations, the 586.2 million cubic feet of water held back by Presa Madín (Madín Dam) in Atizapan de Zaragoza, a suburb of Mexico City, is still used by swimmers and recreational watercraft. In March 2023, some of these brave folks saw something monstrous in the waters.

According to locals, it looked like the Loch Ness Monster. A photograph taken of the supposed creature shows a large black body with a fin surrounded by circular waves.

The city of Atizapan de Zaragoza sent police to investigate.

City officials told the press, "This fact adds to comments from residents of Atizapan, who are sure that a dinosaur lives at the bottom of the dam, which protects the place."

Some locals believe a dinosaur lives at the bottom of the dam to protect the water, others suspect it may be a misidentified giant crocodile, and, given the water's polluted history (although area cities get drinking water from the reservoir, others dump sewage into it), still others think the monster may be nothing but a massive glob of poo.

El Coco

Sleep child, sleep, or else

Coco will come and eat you.

El Coco, Mexico's boogeyman, is terrifying. It sometimes comes as a dragon, or a giant. It sometimes appears as a goblin, or a dark woman with an alligator head. It may emerge from the bushes, or, if you're extremely unlucky, from the dark recesses of your closet, or from under the bed. This insatiable beast glares with fiery red eyes, and it smiles with a mouthful of knifelike teeth.

What does it hunger for? Children.

This mythological monster sleeps all day in a cave, coming into the towns and cities looking for ornery children. When it finds one, el Coco snatches up the child, takes it back to its cave, and swallows them whole.

Really? Gross. As a parent, I can attest to the fact that children are usually dirty.

As all boogeyman tales go, the legend of el Coco was probably constructed to frighten children not to venture into dark, dangerous places. The legend is at least five hundred years old and has been used to describe the child-snatching owls, a creepy old man waiting on the roof with a sack, a dragon, the Devil, or a shape-shifting beast.

What does el Coco mean? The Coconut. Not scary? Pick one up and look at those three holes. Now turn off the lights, and see it in the gray blanket of night. Oh, yeah. That's scary.

The Onza

Long known by local peoples and explorers alike, the Onza is, however, unknown to science. Roughly the size of a mountain lion, although longer and sleeker, it is, however, more vicious than the lion. The first appearance of the Onza in scientific literature came in the 1770 book *Le Système de la Nature* (*The System of Nature*) by philosopher Paul-Henri Thiry, Baron d'Holbach, although he used the name Jean-Baptiste de Mirabaud.

"The Portuguese call it Onza because it is like a Lynx because of its dots. But Hernandes is calling it the Mexican Tiger."

ONZA

Over the years, the Indigenous peoples and Baron d'Holbach have been accused of misidentifying the Mexican Tiger as simply being the local animal, jaguarundi. However, the jaguarundi is relatively small, as wildcats go, growing, at most two and a half feet long, one and a half feet tall, and about fifteen pounds, just a bit larger than a house cat. This doesn't compare to the Onza, comparing it to the mountain lion, at eight feet long, three feet tall, and weighing up to two hundred twenty pounds.

The Onza was reportedly kept in the zoo of Montezuma II. It was mentioned in the sixteenth-century book *The Florentine Codex* by Spanish Franciscan friar Bernardino de Sahagún, in which the friar said it resembled a cougar (mountain lion), but was more aggressive. In 1938, jaguar hunters claim to have shot an Onza and donated its skull to science, although it has since vanished. Sightings since then have become scarce.

Then, what is it? A misidentification? A misplaced big cat? A wild crossbreed?

We'll never know until a specimen is captured and brought to a lab.

Nonhuman Beings

In September 2023, something amazing happened. Journalist Jamie Maussan stood in front of Mexico's Congress and presented boxes containing nonhuman beings.

These two small bodies he claimed were extraterrestrials discovered in Peru six years before. They were mummies, dated to, at most, eighteen hundred years old, according to CBS. They were emaciated, had three fingers on each hand, and looked a bit like Hollywood movie props.

Maussan was all in on them.

"I think there is a clear demonstration that we are dealing with nonhuman specimens that are not related to any other species in our world," he said to Congress. "It's a topic for humanity."

Days later, scientists determined these tiny mummies were "recently manufactured dolls, which have been covered with a mixture of paper and synthetic glue to simulate the presence of skin," according to the *Associated Press*.

Okay, okay. They weren't extraterrestrials. However, in Maussan's defense, they were nonhuman beings.

Michoacán

This Mexica state has a history of strength. Michoacán, in west-central Mexico, with 135 miles of Pacific coastline, is home to the Purépecha (also called the Tarascans), who successfully defended themselves against the warring Aztecs. Michoacán is called many things: the soul of Mexico, the Avocado Capital of the World, and the Birthplace of Carnitas, and the Home to Monarch Butterflies. This state has most of the sites where the monarch butterfly's eastern population goes for the winter. The climate ranges from cool in the mountains to humid and hot on the coast. Much like other Mexican states, Michoacán has volcanos, lots of them. It's also known for a horribly unattractive dog.

Itzcuintlipotzotli

The name is longer than the creature.

Not necessarily a monster, but still a cryptid, the Itzcuintlipotzotli is a hairless humpback dog about the size of a terrier. Hairless dogs of the Xoloitzcuintle breed are common in Mexico. The Itzcuintlipotzotli isn't common anywhere.

Mexican priest Francisco Javier Clavijero wrote of the Itzcuintlipotzotli in the 1780s in his book *Ancient History of Mexico*, describing a medium-sized hairless dog with no neck, the head of a wolf, a large fat nose, a short tail, and a bison-like hump that extended from its shoulders to its flanks.

Although most Itzcuintlipotzotli reports are from Michoacán, in 1843, Frances Calderon de la Barca, wife of the first Spanish minister to Mexico, claimed to have seen a dead Itzcuintlipotzotli in Mexico City. It had been kept as a pet, but its owners put it down when the animal became vicious.

That's it. That's the known history of the Itzcuintlipotzotli, possibly one of the ugliest dogs unknown to science.

The Mermaid in the Mirror of the Gods

Lago de Zirahuén (Lake Zirahuén), in Michoacán's Colonial Highlands, is a 3.75-square-mile blue-water mountain lake surrounded by forests of oak and pine. It's home to the fish Allotoca meeki, a species only found in Lake Zirahuén. However, two types of bass were introduced to the lake in 1933 and have all but exterminated the Allotoca meeki population.

It was the ancient home of the Purépecha Empire, and in their language, the lake's name, *Zirahuen*, means "mirror of the gods," and that mirror is 230 feet deep. In it is a mermaid.

As legend goes, during the Spanish conquest, King Tangaxuan II welcomed the Spaniards until one of the Spanish officers fell in love with the king's daughter, Erendira. Tangaxuan II, as any good father would, didn't approve of her suitor.

So, the Spanish officer kidnapped her and spirited her into the mountains.

Erendira prayed and prayed during her captivity, and the gods of the day and night heard those prayers, and rescued her—by turning her into a mermaid. Erendira's tears were so great, they turned into Lake Zirahuén, where she, supposedly, lives today, only appearing to punish men who wish ill of women.

Morelos

The average temperature for the state of Morelos is 72 degrees Fahrenheit (22.2 degrees Celsius). It's mountainous (the highest point at 18,000 feet), covered in forest, and is close enough to Mexico City to benefit from its economy and services, and far enough away to be private. Is this paradise? Yes. And it has been for a long time. The archeological site Xochicalco was a pivotal city in both the Toltec and Aztec empires. Its ruins include pyramids, an observatory, and ball courts. With an area of 1,911 square miles, Morelos is the second-smallest Mexican state. Not too small, however, to evade the notice of the Aztec Bird King.

Atotolin

The Aztec people had deep respect for nature, and an important symbol for this respect was Atotolin, the Bird King, who represented a balance between the physical and the spiritual. The Aztecs honored Atotolin with hunting rituals.

The bird that embodied the spirit of Atotolin was a water turkey, or, specifically, the white American pelican. They saw it as the ruler of all birds because it didn't live on the shores, but in the heart of the water. They also believed the pelican capable of summoning wind strong enough to sink the boats of fishermen.

Hunters would conduct religious ceremonies while hunting pelicans; this connected the hunter to the Bird King and would give them more prowess during the hunt. However, if the hunt lasted four days, Atotolin would cry and summon winds, and the hunter knew he had failed.

If, or when, the hunter claimed an Atotolin (apparently a difficult bird to hunt), they would split its chest and remove its gizzard, the contents of which would determine their fate. If the gizzard held feathers, or, for some reason, a jade stone, the hunter's future would be successful. If the stomach held a piece of charcoal, their death was imminent.

Killing an Atotolin showed the hunter had a deep connection with the spirit realm, which I guess is necessary to find a bird with a gem or coal in their stomach. Yes, some birds swallow gravel to allow a gizzard to better break down food, but what birds eat jade and charcoal?

Okay, okay. I'm not a Bird King, so whatever.

Cipactli

In the beginning…

In Aztec mythology, water covered the world. It took a god to bring about land. For the Aztecs, this god was a huge creature named Cipactli. And what a creature it was.

Cipactli was an enormous sea demon, part fish, part crocodile, part toad—and it was always hungry. As other gods created creatures of the sea and land, Cipactli ate them with its eighteen mouths. Angered that this demon would eat everything they created, the other gods fought it, Cipactli eating a foot of the god Tezcatlipoca before it was defeated.

Then the gods pulled the demon in the direction of the four winds and used the pieces of its body to finish building the earth and the sky.

Geez. Creation is brutal.

Azcatl

Not quite *Empire of the Ants*, but close enough. (If you didn't get that reference, you're not into shlock movies involving enormous radioactive insects.)

In Aztec folklore, Azcatl were ants the size of three jaguars. At maximum size, that's potentially a twenty-four-foot-long, nine-hundred-pound ant. For perspective, that's as long as a box truck, and as heavy as a grand piano.

The Azcatl's job in the world of the Aztecs was to protect the Place of Herons, a number of lakes with a rock in their center. A cactus grew from the rock, and on the cactus, an eagle forever devoured a serpent. If anyone, or anything, threatened the Place of Herons, Azcatl would hypnotize them and protect the site.

Nayarit

Nayarit ranks as the twenty-third-largest Mexican state, but since there are thirty-one states, that's not saying a lot (although it's nearly six times the size of Morelos). It's on the Pacific coast, boasting the 190-mile Riviera Nayarit, a beach that stretches the western length of the state, and attracts nearly 3 million tourists per year. The rest of the state is covered by mountains, forests, and cities with beautiful colonial homes, and San Blas's Fuerte de San Basilio, a fort built to deter pirates. Mexcaltitlán, a man-made island built off Laguna Grande de Mexcaltitlán, may be the birthplace of the Aztecs. Historians are still arguing that one. Nayarit is also home to a mystery cat.

Nayarit Ruffled Cat

Deep in the mountains of Nayarit lives (well, could live) a wildcat unknown to science.

NAYARIT RUFFLED CAT

This cat, the Nayarit Ruffled Cat, is between four and seven feet long with brown fur and a large mane and light brown stripes on its flanks. It has long legs, big paws, and a tail just shorter than two feet long. One of the cat's most recognizable features is its prominent fangs.

Nayarit has several cat species, including the jaguar, ocelot, bobcat, jaguarundi, mountain lion, margay, and the domestic cat. None of which fit the description of the Ruffled Cat.

Biologist and cryptozoologist Ivan T. Sanderson bought what were purported to be the hides of the Ruffled Cat in a mountain village. Accounts of what happened to the hides differ. One claims the hides were lost in a flood before they could be properly examined, the other is that the government of Belize accidentally destroyed them.

The Nayarit Ruffled Cat is believed to be anything from an unknown species, to a misidentification of a known animal, to a surviving breeding population of saber-toothed cat that roamed the Americas until ten thousand years ago.

The Seven-Headed Snake

The symbol of a snake with seven heads is represented across the world: Lotan of Ugaritic (early Semitic legend), the Greek Hydra of Lerna killed by Heracles, the Sumerian Mušmaḫḫū, the Albanian Kucedra, and the seven-headed dragon (Satan) in the Christian *Book of Revelation*.

It is also represented in Aztec lore.

In the town of Jala—a Magical Town—is surrounded by cliffs, and hills, and is loomed over by the active volcano el Ceboruco.

It is called a Magical Town because of la Quema de la Sierpe (the Burning of the Serpent).

In local legend, a great seven-headed serpent would rise during the summer rainy season and destroy the corn crops. In the 1800s, Jala villagers began a festival to ward off the enormous serpent. They spread flowers throughout town, played music, ate lots and lots of the area's strangely huge corn, then killed the serpent.

They don't actually kill the serpent, because it's represented by townsfolk in costume, in which archers symbolically destroy the monster on a hill outside town. The death is commemorated with fireworks.

Part of this ceremony involves the Virgin of the Nativity—the Virgin Mary of Catholicism. Yes, these go together. Nayarit is 88.1 percent Catholic. In art, the Virgin Mary is often depicted standing on the neck of a snake. This is because of Genesis 3:15: "I will put enmities between thee and the woman, and thy seed and her seed: she shall crush thy head, and thou shalt lie in wait for her heel."

The Catholic interpretation of this verse is that the Mother of Christ will cause the defeat of Satan.

The Bible didn't mention corn in this passage, but it doesn't matter; la Quema de la Sierpe is also a corn festival, and there's a good reason for it. Everywhere but Jala, an average cornstalk grows from between five and twelve feet tall, and its ears are between six and a half and seven and a half inches long, but in the fields around Jala, corn plants grow up to twenty feet tall and produce ears twenty inches long.

There is another seven-headed serpent woman in Aztec mythology, and there's a distinct difference between the Jala tradition and the Aztec legend of *Chicomecóatl* (Seven Serpent). Chicomecóatl is a goddess of food, especially corn, and is depicted as a flower-bearing girl, a mother using the sun as a shield, or a woman with seven snakes for a head. Chicomecóatl is the goddess of food and drink and is claimed to have created the tortilla.

Why the villagers of Jala would destroy this life-sustaining goddess is beyond me.

Regardless, la Quema de la Sierpe is held every September 8, the day after my wedding anniversary—don't forget to send corn.

Nuevo León

The state of Nuevo León is named after the Kingdom of León, a nation (910–1230 CE) in the northwest of the Iberian Peninsula, taking up parts of what is today Spain and Portugal. Nuevo León is in northeast Mexico and shares a nine-mile border with Texas; the rest of the state is surrounded by Coahuila, San Luis Potosí, Tamaulipas, and Zacatecas. Its land consists of mountains and plains, and its major products are cotton, sugarcane, wheat, corn, and vegetables. Wildlife includes pit vipers, antelopes, black bears, and, apparently, goblins.

Monterrey Goblin

The capital of Nuevo León, Monterrey, is the third-largest city in Mexico with a population of 1.1 million (a far cry from Mexico City's 22.5 million, and half that of Tijuana's 2.1 million, but still). It's a hub of manufacturing and services like the hotel and restaurant industries. The city lies in the Sierra Madre Oriental Mountain Range, surrounded by natural beauty.

On a farm on the outskirts of this metropolitan city, a security camera captured something out of place crossing the road on

MONTERREY GOBLIN

December 1, 2021—and, when the video was posted online a month later, it caught people's attention.

The video was of a dark, unidentified, bipedal creature about a foot tall crossing a dirt road. The video shows the entity entering the road cautiously, scampering across, then seeming to relax as it slows before disappearing into the foliage on the opposite side.

The creature is definitely walking on two legs, and there are no known animals of that size, and behavior, that are indigenous to Nuevo León. There are, however, plenty of stories of diminutive humanoid entities that live in farmland throughout Mexico.

So, a goblin? Why not?

Quetzalcoatl—Part 2—Maybe?

The Aztecs didn't expand as far north as Nuevo León, but their chief god may have.

The Alazapa, Borrado, Coahuiltecan, and Guachichile peoples have inhabited Nuevo León for centuries, but during Spanish colonization, a story emerged from a Spanish explorer, Alonso de León, that places the key figure in Aztec mythology squarely in Nuevo León. Quetzalcoatl.

Between Monterrey and what is now the border of the United States, he crossed the San Juan River—the major river in Nuevo León—and found a bottomless spring that, although the source of the water could not be found, the pool never grew, and it never diminished. The local peoples said they were blessed from it.

The story of the pool involved a man from the south, claiming to be a god, teaching the local tribes agriculture and showing them the beauty of the magical pool before he departed.

Another story tells an ugly old man appeared after then god left, calling the god a liar, and a demon, before he left as well. When the god returned, the people didn't believe his words, and the god never came back. The ugly man never returned either.

Alonso de León speculated the first visitor may have actually been Quetzalcoatl come to bless the people, and that the ugly one was an angry Spaniard who wanted nothing more than to ruin joy for the Native peoples.

The Monterrey Bird Man

The mountain Cerro de la Silla is at the eastern end of Monterrey and rises 5,970 feet above the city. Locals say that sometimes they see the Monterrey Bird Man on the mountainside. Oh, and not just there, sometimes it descends upon the city at night and lands on the roofs of residential homes.

Witnesses have reported the Bird Man, first seen in the 1980s, is a black, glowing-eyed, man-sized creature with a fifteen-foot wingspan.

Vampire, space alien, extant species of pterodactyl—people have speculated its identity as nearly anything. Some have claimed it wears a pentagram, marking it as something demonic. Regardless, homeowners have reported after a night of terror with the Bird Man stomping across their rooftops, the next day they discovered claw marks.

Oaxaca

There are sixteen significant Indigenous cultures in the state of Oaxaca, which have survived into the modern world mostly due to isolation. Three mountain ranges cover 82 percent of the state, but two, Sierra Madre del Sur and Sierra Madre del Norte, run through its entirety, making for a rugged land to travel (there are a total of 2,786 mountains in Oaxaca). The mountains and valleys, coupled with 370 miles of Pacific coastline, make Oaxaca perfect for cultural preservation, and, of course, tourism. It has dry tropical forests, temperate forests, and plenty of farmland for more than thirty different types of agave used to make mezcal. It is one of the top biologically diverse states in Mexico (having more species of amphibians and reptiles than any other state), and, you guessed it, it's also home to creepy-crawlies.

Camazotz

Oaxaca being home to vampire bats is bad enough. Sure, they suck blood; sure, a lot of them carry rabies; but the body of a common vampire bat is only two inches long, it has a wingspan around eight inches, and it weighs an ounce (the weight of a pencil, or a slice of bread). But the Camazotz? Compared to a common vampire bat, it's basically a Kaiju monster.

Although Camazotz is a Maya deity, the Zapotec peoples of Oaxaca built a cult around this creature that lasted in the open until Spanish colonization, and, who knows? It could still exist somewhere in the state.

Camazotz—"Death Bat," to the Maya—is linked to darkness (duh) and death and requires a sacrifice. According to the *Manzanillo Sun*, Camazotz is the "death bat god of the underworld." It was a leaf-nosed bat as big as a man, with a wingspan of around eight feet. In Maya depictions, the creature had a knife in one hand and a human heart (still beating, I suppose) in the other. It was known to snatch its victims from the ground and rip off their heads as the monster flew away. Lovely.

Some speculate the god (usually goddess) Camazotz was modeled after *Desmodus draculae*, a giant vampire bat (okay, it was 30 percent larger than today's tiny version, but that's still big) that became extinct between the mid-1600s and the early 1800s.

Mometzcopinqui

Women born on Ce-Ehecatl (which means "One-wind" or "One-rain") have no choice in their life. They're going to become a *bruja* (a witch).

It's not their fault; the gods chose the fate of those women on the day they were born.

A fireball at night in the hills denotes the presence of a Mometzcopinqui. These bruja can remove their arms and legs and replace them with the wings and legs of a turkey. To protect their appearance of humanity, the Mometzcopinqui hide their limbs beneath their homes to use later.

When transformed, they fly in search of their prey, which are babies, then feed on the blood of the infants. When the bruja returns to its home, it releases the blood so it can feed for days.

The Mometzcopinqui legend has continued until today, the turkey still being an important part of Mexican culture.

Oaxaca Alien Werewolf

It's enough to have a Nagual (werewolf) in your midst, but to have one that is a space alien? That's another matter.

In late November 2023, Oaxaca ranch owner Román Ochoa told the *US Sun* a man changed into a werewolf before his eyes. He claimed it was a Nagual (a witch that can shape-shift into a spirit animal), although he also said the man got this werewolf ability from a space alien.

Ochoa saw this alien more than once, but this time, he got a photograph.

The wolflike human monster glared at the farmer as he took the picture, the flash giving it white, glowing eyes. Its front paws looked like hands, its rear paws like human feet.

No one on the ranch was harmed by the creature, and it slunk off into the undergrowth and disappeared.

A comment on a local news read, "Stop drinking tequila so early, excess is not good for your health."

Puebla

The state of Puebla is high. Totally high. The average elevation ranges from between 5,000 and 8,000 feet; the highest point is the top of the dormant volcano Iztaccíhuatl at 17,159 feet. Told you. It's also home to the largest pyramid in the world—the Great Pyramid of Cholula. Although at 217 feet tall, this pyramid isn't the tallest (that's the Great Pyramid of Giza at 454 feet), but it's the largest by volume at 157.1 million cubic feet (Giza's is 92 million cubic feet). The state is also known as the home of mole poblano sauce, products made from onyx, and the Little Devil.

The Little Devil of San Miguelito

The Church of San Miguelito is one of the oldest—if not the oldest—church in the city of Cholula, itself the oldest (or one of the oldest) continuously lived-in cities in the Americas. The church, dedicated to the archangel St. Michael, once had a statue of Michael near the altar, and a red Devil at his feet.

The Devil statue—that appeared outside the church doors one day—represented the victory of the archangel over evil. However, when parishioners came in to pray to the statue of St. Michael, they would also pray (quieter) to the statue of the little Devil, just to hedge their bets. Who really knew which one was listening?

This practice outraged more straitlaced church members who believed this empowered Satan; they blamed the people praying to the little Devil on any evils that befell Cholula. When locals began reporting seeing the carving of the little Devil walking around town at night, people would storm the church—and often find the little Devil missing, only to reappear in his usual place the next morning.

So many accidents occurred on the road running by the chapel, the road was eventually called la Curva del Diablo (the Devil's Curve).

Both statues are now gone, as are blessings, and curses, and la Curva del Diablo has gone silent.

The Woman-Dog of Los Sapos

A dog roams the streets in a bar and restaurant district in the city of Puebla, but sometimes it's a young woman. If you're drunk in the Los Sapos area, beware, or punishment for your merriment will come swift.

Women who've imbibed a bit too much alcohol may be approached by a puppy, wagging its tail, begging for attention. If the woman bends to pet this puppy, the canine changes, growing suddenly into a huge snarling black dog that bites and scratches the drunken woman until she stumbles home.

Men may meet a beautiful woman who flirts with them, coaxing them to follow her into a secluded spot. If the man does and attempts to kiss the woman, when they open their eyes again, they see they've actually kissed a skeleton.

Charming legend. I think I'll have another beer.

WOMAN-DOG
OF LOS SAPOS

Cuatlacas

Latin America has a variety of Bigfoot-like creatures: the Quinametzin of Campeche, Sismite of Belize, the el Sisemite of Guatemala, and, although, unlike some of these, the Cuatlacas doesn't have its feet on backward, these creatures are all big, humanoid, and hairy. In Puebla, the Bigfoot is the Cuatlacas.

The Cuatlacas stands between seven and nine feet tall, is covered in thick dark hair, and is shaped like a man. It's also incredibly strong, and incredibly angry. Not the best combination.

Much of Puebla is covered in mountains, valleys, riverbeds, and forests—old-growth oak and pine forests. The perfect habitat for the Cuatlacas.

The Cuatlacas is thought of as a guardian of the forest, appearing much like its North American cousin in size, fur, and facial features of a man, leaving only huge, bare footprints in its wake. However, it behaves much like the territorial Australian Yowie, so don't wander into its land, or it will warn you off with howls, thrown rocks, and it apparently once destroyed a man's truck.

This creature has been part of the mythology of Puebla for at least two thousand years. I'm not one to argue.

Querétaro

An aqueduct runs through Santiago de Querétaro, the capital of the state of Querétaro. Built of pink stone between 1720 and the 1730s, this aqueduct draws the water of local springs to the city and has become a symbol of Querétaro. It is five and a half miles long, about forty-two hundred feet of it held by seventy-four arches, some reaching ninety-four feet tall. A former part of the Aztec empire, Querétaro was taken over by Spain in 1531. The state produces cotton, textiles, heavy machinery, and agricultural goods. Space aliens and demons seem to like it, too.

The Gray Aliens of Peña de Bernal

The monolith Peña de Bernal stands 1,421 feet above San Sebastián Bernal, a small town with a population of just around three thousand people. With the look of Devil's Tower in Wyoming (although about two hundred feet taller), Peña de Bernal (Rock of Bernal) is the second-largest monolith of its kind in the world, only behind Australia's famous Uluru Rock (formerly Ayers Rock). All sacred sights to Indigenous peoples.

In the 1977 movie by Steven Spielberg, *Close Encounters of the Third Kind*, Devil's Tower plays host to space aliens. Uluru Rock, and the Rock of Bernal, do so—in real life. Hmm.

Spirits are said to inhabit Peña de Bernal, as do Chaneques, the fae folk of Aztec legends. Sightings of Chaneques continue to this day, although modern visitors to this sacred site describe them with a more modern term—*the Grays*.

Diminutive gray humanoids with large heads, holes for a nose, a line for a mouth, and enormous, black, almond-shaped eyes are sometimes encountered by visitors to the Rock. Stories of the Grays—Zeta Reticulans, Roswell Grays, etc.—have been around since the late 1950s/early 1960s, becoming mainstream with the publication of Whitley Strieber's 1987 book *Communion: A True Story*. The picture of the gray alien on the cover scared the hell out of readers.

The fact that people have reported encounters with Grays on Peña de Bernal isn't just fascinating, it's expected. Not only is Peña de Bernal a haven for UFO/UAP sightings, the Chaneques connection corresponds to the abduction behavior of fae folk worldround.

But Peña de Bernal isn't only an extraterrestrial hot spot; visitors have reported meeting the Count of St. Germain there.

Oh, the Count of St. Germain. A self-proclaimed nobleman of possible European descent who began to appear in royal circles around Europe in the 1600s. He never aged, and claimed to have been Plato, Merlin, St. Joseph, Columbus, and/or was born in Atlantis.

He has apparently appeared in many centuries all over the planet, impressing people with his wealth and knowledge, educating them with his wisdom, then disappearing. Peña de Bernal may be one of the places he's visited.

Regardless, if you're ever in the area, stop and visit. The monument and the surrounding areas are beautiful. And, who knows, you may be abducted by space aliens, or meet an immortal saint.

The Devil in Querétaro

The story, as many stories with a bad ending do, begins with greed.

In the 1600s, a moneylender in Santiago de Querétaro named Bartolo de Sardanetta was in love with his sister (*ewwww*). To build himself into a desirable man, he called on the Devil, and the Devil appeared. Agreeing to trade his soul for money, the Devil smiled and granted Bartolo de Sardanetta his wish.

His sister, not into the whole incest thing, rejected him, so he murdered her. When de Sardanetta eventually died, the Devil collected his soul in a puff of sulfur.

This is only the beginning of the Devil in Querétaro.

Also in the 1600s, a spirit pushed a girl into the Lerma River, and, although a passersby rescued her, she wasn't the same. Satan possessed the girl. She wasn't the last. She was one of the *Endemoniadas de Queretaro* (Possessed of Querétaro).

As demonic possessions grew in the town, in the early 1690s, Colegio de la Santa Cruz de Querétaro (College of the Holy Cross of Querétaro) chose to combat the Devil. Their tactics focused on women.

Whenever a woman showed a symptom of possession, the friars would perform an exorcism; however, not all were successful. Some made the women behave "worse," some of them becoming pregnant by Satan himself. Eventually, the townspeople tired of this and blamed the friars of impregnating the women themselves, using the Devil as a cover.

Then there was Claudia Mijangos, the Hyena of Querétaro.

Claudia Mijangos was born in 1956, wealthy, and as she grew, beautiful, as a teen achieving the honor of becoming the beauty queen of her hometown of Mazatlán in the state of Sinaloa. After graduating school, she married Alfredo Castaños Gutiérrez, and they had three children. In the late 1980s, Claudia and Alfredo separated, partly due to Claudia's increasing mental instability.

In the ensuing months, visions plagued her. Demons and angels, Devils and saints, all visited her, telling her what to do. On April 24, 1989, demonic voices drove her from sleep, then sent her to her children. She stabbed them all to death while they slept.

Although doctors diagnosed her with schizoaffective disorder and temporal lobe epilepsy, she maintained the demon that possessed her forced her to kill her children. She finished serving her prison sentence in 2019 but remains under psychiatric care.

Is Santiago de Querétaro cursed by demons?

Who knows? The closest answer to this question comes from the Catholic Church, which opened an exorcism chapel in the city in 2010, according to the BBC. This directive specifically calls for the priest to rule out epilepsy, schizophrenia, or other psychological issues before an exorcism is performed.

Quintana Roo

The youngest and easternmost state of Mexico, Quintana Roo's tourist-filled beaches reach into the Caribbean Sea and the Gulf of Mexico. The most famous city on the state's Riviera Maya (Maya Riviera) is Cancún, its most famous island Cozumel, but Quintana Roo isn't just beaches and drinks in cups with little umbrellas. The country is covered in lush forests, watery sinkholes, and scattered Maya ruins. The people who inhabit Quintana Roo are descendants of the Maya who ran the non-Indigenous people off the Yucatán Peninsula in the Caste War from 1847 to 1853. They came back. The state has also been known as a home for sea monsters.

Ah Puch

The god of disaster, death, and darkness, Ah Puch rules the Maya underworld, Xibalba, the place of fright. This being is depicted as a skeleton with the head of an owl, although sometimes it appears as a bloated corpse wearing bells. This creature's job is to keep souls confined to Xibalba.

However, it also goes out to hunt for fresh ones.

This skeletal figure has been seen lurking in the shadows as it waits outside the houses of the sick, waiting for them to die. When a family member died, Maya would traditionally mourn loudly to frighten Ah Puch away before the monster could steal their loved one's soul. People could tell when Ah Puch was around due to its putrid stench.

AH PUCH

Ah Puch was also a *Wayob* (meaning "sleep"), a being that transformed into an animal while asleep and set out to harm people. Some of these Wayobs are Nagual (also spelled *Nahual*), who transform into a bear or a wolf to capture souls and transport them to Xibalba. As a Wayob, Ah Puch is a putrid, rotting god of death (near the coast, that death is usually in water), but also a god of childbirth and regeneration.

Che Uinic

Che Uinic is on one hand terrifying, on the other hand weird. This giant man of the forests from Maya legends is a cave dweller covered in red hair; its deep voice has been said to stun people who hear it, according to the news source *Universal*.

The giant's favorite food is...yeah, people.

The giant cannibal is the scary part. Here's the weird part.

Che Uinic has no bones. Not one. And its feet are upside down, so when he walks, he walks with the help of a log (he's a giant, remember. That's a cane to him). Then, of course, for its victims, running away isn't too hard. This giant can only feed by hiding and waiting on unlucky, and unobservant, passersby.

According to legend, when confronted by Che Uinic, if you sing, dance, and juggle, it will collapse in laughter, which is easy for it because of its lack of a skeleton.

San Luis Potosí

This state in eastern-central Mexico is known for silver and gold. At least it was. During the Spanish conquest, miners discovered these precious metals, along with copper, bismuth, and zinc in San Luis Potosí. One gold load produced about $60 million between 1900 and 1910. San Luis Potosí is on the Mexican Plateau with an average elevation of around 6,000 feet. The jade Tampaon River flows through the tropical Veracruz moist broadleaf forest in the eastern part of the state. One of the famous residents of San Luis Potosí was Francisco González Bocanegra, who wrote the "Himno Nacional Mexicano" ("Mexican National Anthem"). Then, there are giants.

The Giant Thin Man

The city of Ciudad Valles lies near the eastern edge of San Luis Potosí. On a day in late May 2020, Mr. and Mrs. Antonia told *Sol de San Luis* they encountered a tall, thin entity from legends.

"I was with my husband playing with my child when we heard a noise, like an explosion, then we looked at the hill, where we thought the noise was coming from, we thought it was a landslide," Mrs. Antonia told the newspaper. But the sound didn't come from a landline, it came from a living being that looked "like a giant."

Other citizens of the area saw the creature too. Residents of the mountains, especially, have seen the thin giant. Some believe it to be Bigfoot. The older people in the community, *Sol de San Luis* reports, should be more likely to believe in giants, but do not; however, they have discussed what the discussion of giants, the COVID-19 pandemic, and a recent spate of UFO sightings means to the world in general.

A photograph taken of the creature is, as most photographs of cryptids, fuzzy and indistinct.

Santiago River Frogman

Nineteen-year-old Francisco Estrada Acosta was hunting small game with a slingshot when he came face-to-face with an impossible creature near San Jose in San Luis Potosí on February 12, 1965.

SANTIAGO
RIVER FROGMAN

According to *Catalogue of Humanoid Cases 1965–2006* by Albert Rosales, Acosta saw a "tall scaly creature that extended a cold hand as an attempt to possibly greet the witness." He knew the hand was cold because the creature touched him with it. Terrified by the appearance of a monster, Acosta ran.

Francisco Estrada Acosta had encountered the Santiago River Frogman.

He said the creature had "phosphorescent eyes, a large toad-like mouth that suddenly appeared next to him on the road. Its hands were like 'flippers.'" It was more than six feet tall. As he ran, he turned to look behind him and saw what looked like wings extend from the Frogman's back, according to the book *Contact: Mexico: History of the UFO Phenomenon* by Luis Ramirez Reyes.

Acosta didn't wait around to discover if the beast could fly; he ran faster.

Unfortunately, for the sake of study, Acosta seems to be the only witness to the Santiago River Frogman.

Sinaloa

Sinaloa, in northwestern Mexico, is known as Mexico's Breadbasket for producing 40 percent of the nation's food, growing rice, chickpeas, wheat, tobacco, cotton, and sugarcane; it also has the highest number of aquaculture farmers and produces the popular pilsner beer Pacífico. It has four hundred miles of coastland on the Sea of Cortez (the Gulf of California) to the west, and the Sierra Madre Occidental Mountain Range to the east. But the state is not all beauty and farming; there's also the man with low pants.

El Calzonudo

In the town of Rosario, a strange figure the locals call *el Calzonudo* (the Naked Trousers) has been known to stalk townsfolk. For decades, a figure dressed in a white suit of manta (traditional material made from wild cotton) has been seen in fields outside town, pants to his knees.

According to a November 2019 article in the *Mexican Post*, a local man, Felipe Ruiz, saw el Calzonudo himself. Ruiz was walking to work at an area called *Palo Quemado* (Burnt Stick) when he saw the figure in the white manta suit.

"I see a man on the road," he told the newspaper. "The way he was dressed, I had no doubt that it was the Calzonudo."

Momentarily stunned, Ruiz could only stand and watch as the figure strolled closer.

"Fear came to me when I looked at him," Ruiz said. "All I did was run, trying to get to town because I felt that this man wanted to take me with him."

Ruiz climbed a fence to another field and ran some more. When he turned to look behind him, el Calzonudo (with his pants around his knees, of course) walked through the fence "as if it were wind," Ruiz told the newspaper. "I ran even faster after seeing that."

Although Ruiz ran, and el Calzonudo only walked, the white-suited entity was closing the gap between them.

Ruiz reached a stream and slogged through the fast-moving water. When he reached the other side, he turned toward the opposite side, and el Calzonudo was gone.

Although he kept the story to himself for a few days, he finally spoke to people around town who confirmed that, yes, it had been el Calzonudo chasing him.

The Sinaloa Vampire

Canóvanas, Puerto Rico, resident Madelyne Tolentino was the first person to report an encounter with a creature soon to be called el Chupacabra. This was in August 1995. Less than a year later, the first report in Mexico came from a village in Sinaloa. Soon, the word would be out—the Chupacabra has come to Mexico.

The following information is from a May 19, 1996, article in the *Los Angeles Times*.

The farmer woke to find twenty-four of his sheep lying dead on the ground, every drop of blood drained from two puncture holes in the neck of each one. Then, a creature that looked like an enormous bat flew from the corral. The farmer called the authorities, and the press weren't far behind.

What killed this farmer's sheep? A monster bat? A vampire? A space alien?

Sinaloa's chief of civil protection, Desiderio Aguilar, told the *Times* this attack, along with new reports that kept coming in, sent a wave of terror across the state.

"This created a great panic," he said. "Suddenly, normal deaths of chickens and goats and sheep are all being blamed on the goat sucker...Mothers have quit sending their children to schools for fear they could be attacked on the way. Farmers who used to start work at 4:00 a.m. to beat the heat aren't leaving their homes until well after daybreak."

The federal government sent agents to the area to quash talks of a monster, blaming the attacks on a natural predator, like a big cat, or coyote.

It didn't help. Rumors persisted. More sheep died in this manner, along with goats, chickens, turkeys, and rabbits. A worker on the ranch of the governor of nearby Durango claimed to have seen a "yard-high dinosaur kind of thing with fangs, bulging eyes, bat wings, a needled spine, and kangaroo legs."

Parents across the country were afraid for their children, keeping them inside at night; people in small towns attacked bats in caves with torches.

In Sinaloa, biologists, experts from the local zoo, and a SWAT team gathered to collect evidence, and even went as far as putting sheep in the corral from the initial attack, waiting for the monster to come.

It didn't.

Wild dogs, however, appeared and attacked the sheep, "leaving the same marks found on the first dead sheep," Aguilar said. "They captured the dogs and showed them to the townspeople. They went to the other municipalities where there had been attacks and came up with the same results."

Did the people of Sinaloa believe the scientists? No. Me neither.

Sonora

Bordering the above Sinaloa to the south, Chihuahua to the east, Baja California to the northwest, and the American states of Arizona and New Mexico to the north, Sonora—the second-largest Mexican state—takes up 71,403 square miles of mountains, desert, plains, and 507 miles of beach on the Sea of Cortez. Although mining and industry make up a sizable part of the state's economy, cattle ranching makes up the rest. Moo. Wildlife includes wolves, mule deer, jaguars, coyotes, and lots and lots of reptiles—even reptilian humanoids.

Reptilians

San Luis Río Colorado is a city of 200,000 people, close enough to the US border and Yuma, Arizona, the smaller Yuma could be considered a suburb. It's hot there. This city in the desert reached a record 125.6 degrees Fahrenheit on June 25, 1951. The perfect environment for snakes (oh, so many snakes), desert iguanas, geckos, horned lizards, and…

Well. What do you know.

Local farmers have reported a six-foot-tall, manlike lizard lurking on their properties, stealing turkeys and chickens, leaping great distances when confronted. The lizard men come alone, or in small groups, but flee before anyone gets off a shot.

What are they? A race of dinosaur people? Uppity iguanas?

Extraterrestrials, of course.

According to the paper "They Are Among Us: Mimicry and Disappearance Against Reptilians in Mexico" by Andrea Murillo of the College of Michoacán, "Reptilians, aliens with a humanoid and reptilian appearance that live in Mexico and on Earth, pose as humans to dominate them."

And they're in Sonora.

These creatures have been reported to have the ability to not only camouflage themselves, but to transform themselves into human beings who have relationships and work jobs.

The reptilian is, frankly, the most terrifying creature written about in this book.

Alien Skulls

In 1999, outside the small town of Ónavas (population fewer than four hundred people), workers digging an irrigation channel disinterred a skeleton that was human…ish? The skeleton was fine, but the skull? It didn't look exactly normal.

The diggers contacted the authorities, and, after more digging, archaeologists found a cemetery holding twenty-five bodies with elongated skulls, according to Fox31, KDVR-TV, in Denver, Colorado. The skulls were about one thousand years old. Seventeen were from between five months and sixteen years of age, the other eight were adults. The teeth of five skulls were filed to points.

Head binding, which forces the growing skull into a cone shape, is well known, and was practiced worldwide. The scientists said that's what happened in the plot outside Ónavas.

Not everyone agreed.

Were they a different species of human? Or were they human–space alien hybrids? DNA tests determined the skeletons were human, although the shape of the heads weren't necessarily caused by binding.

What are they?

Dead. They're dead.

Tabasco

Nestled between the Gulf of Mexico and Guatemala at the base of the Yucatán Peninsula is the state of Tabasco. Tabasco is beautiful, covered in jungles and lakes, with white sandy beaches, the famous giant Olmec heads, and Maya ruins. Farmers grow cacao, corn, coconuts, bananas, papayas, and, of course, the red pepper that bears the state's name. The Olmec civilization (1200 to 800 BCE) began in Tabasco, building the first pyramid in Mesoamerica at the city of La Venta, a place of culture, astronomy, and blood sacrifices. Because of the thick forests, Tabasco is home to a great number of jaguars (the cat, not the car), and where there are jaguars, there are were-jaguars.

Were-Jaguars

The Olmecs were the first advanced civilization in Mesoamerica, not only building the aforementioned pyramids and cities, but creating the region's first numerical and writing systems. They had a complex history and religious structure, and in that was a half-human/half-cat monster—the were-jaguar.

Many carvings of this creature depict it in different stages of transforming from a man into a man-cat, the heads in some more jaguar-like, the eyes almond-shaped, its teeth sharp and vicious. These carvings have been found across the ancient Olmec civilization, such as the pre-Columbian city La Venta.

People with PhD behind their names credit these depictions of the were-jaguar on figurines and altars as symbols of strength or fertility. Some claim these carvings represent what amounts to a costume change where shamans dressed as an animal to gain their power.

What they don't agree with is the stories of Indigenous peoples who claim to be able to transform into an animal. From the Navajo skinwalkers, to Norse berserkers, to the Odiyan clan of Kerala, India, people across the globe have long claimed to have this ability. Why not Olmec were-jaguars? *Merow.*

Mazateupa Goblins

In the 1960s, on a ranch named la Sabana outside the small town of Mazateupa, the family who lived there raised turkeys, ducks, pigs, and chickens. Lots, and lots, of chickens.

Then, the chickens got weird. Really weird.

They walked strangely, their necks bent strangely, and they sometimes fell down for no real reason. Weird.

When the sun set, the chickens no longer exhibited this odd behavior. When word got to town, the townsfolk drew one conclusion. According to the *Herald of Tabasco*, they believed the birds to be possessed by demons. The family simply called them "los pollos cachurecos"—the deformed chickens.

MAZATEUPA GOBLINS

Worried about his poultry, the father of the family went to the local brujo, who told them demons didn't possess their chickens, it was goblins who didn't like the family living on their land. The family moved, and their troubles did not move with them.

Dtundtuncan

The Dtundtuncan doesn't have eyes, and is missing a leg, but this bird doesn't need them.

It's not pretty, it's not melodious (at least, in some stories), and it has no soul. When the sun sinks below the horizon, and children snooze away at night, the Dtundtuncan flies to their houses, creeps into their rooms, and blows the winds of death into them.

This Maya creature is not only a terror to families, it takes souls to the afterlife and guards sacred places. It appears as a shadow in the sky, silently descending to speak to travelers in the dark, hypnotizing them with its voice.

This giant one-legged bird is thought to be the physical form of a demon sent to Earth to wreak havoc. Other stories claim it to be a Maya shaman who lived to heal women and children, who fell in love with a woman who didn't love him. Years passed, and his resentment grew.

When the woman married and became pregnant with the future prince of the kingdom, this shaman sang to the baby, and sucked out its breath. Other shamans banded to curse this man, and he became the blind evil bird known as the Dtundtuncan.

Tamaulipas

In the northeast corner of Mexico lies Tamaulipas, the sixth-biggest state by size and boasts fertile cropland—the most irrigated in any Mexican state. Its east coast is the Gulf of Mexico, to the north is Texas, Nuevo León to the west, and it is touched by

Veracruz and San Luis Potosí to the south. Most of Mexico's oil and natural gas comes from Tamaulipas. The state is known for its theater groups and is home to Teatro de la Reforma (the Opera Theater), which was built in 1865, and in 1904 hosted the first performance of the Mexican National Anthem by its composer Don Jaime Nuño (as stated earlier, Francisco González Bocanegra wrote the lyrics). Now, how are you with vampires?

The Vampire Woman of the San Juan Bosco Church

The city of Ciudad Madero, part of the Tampico metropolitan area, lies on the Gulf Coast, and, in 2024, GOBankingRates named it the safest city to live comfortably in Mexico. That doesn't mean it's completely safe. There's still crime, and, oh yeah, a vampire.

The following tale was reported in the *Expreso Press*.

In the 1960s, at the San Juan Bosco Church in the Arbol Grande neighborhood of Ciudad Madero, Father Ignacio Rosiles Namorado worked hard to restore the church building. Seeing this man of God's dedication, a socialite wanted to save the building, too, and told him since she gave so much of his life to the church, before she died, she wanted her remains to remain there.

The woman died of old age and was buried in a Salamanca, Guanajuato, cemetery in 1973. The priest, remembering the woman's wishes, asked her family if he could disinter her body and place her to rest inside the San Juan Bosco Church. Wanting the woman's remains to be forever placed in the church she loved, they agreed.

Then the vampire story began.

When diggers reached her coffin, they found her body as it was buried. No decay, her hair and nails had grown (common misconception to a normal part of the process), but so had her teeth—

the woman had fangs. She became forever known as the Vampire Woman.

As per the priest's wishes, the woman's body was moved to the church, where he held mass for her. As parishioners crowded around the casket, they knocked it to the floor, and witnessed the Vampire Woman with their own eyes.

Many of the churchgoers ran home to gather equipment they would use to set the church on fire. When they stormed back inside the church, they didn't see the Vampire Woman, and suspected her hidden in another part of the building.

Members of the Mexican Army supposedly arrived to stop the panic-driven mob.

Panic-driven indeed. Some people claim to have seen the Vampire Woman fly from the church and disappear into the night. Others said they witnessed her attacking children and pets. Those stories continued over the years, keeping young children inside when the sun sank below the horizon.

The Dog Girl

A girl began appearing on security cameras in Tamaulipas in 2009. She was naked. She was fast. She ran on all fours.

What?

Reported on Portland Oregon's Z100 FM in December 2023, the Dog Girl has never taken a break. A 2020 video accompanied the article; the girl (really? A girl?) ran smoothly, effortlessly, as if she were used to running like an animal.

People following the story have speculated the girl to have been about fourteen years old in the 2009 video, putting her close to thirty years old today.

Medical experts have weighed in and suggested the Dog Girl (*Niña Perra*) has Uner Tan syndrome, a genetic disorder that causes those suffering from it to walk on all fours, although never like this.

Whatever the Dog Girl is, she's mysterious, fast, and buck naked.

Tlaxcala

At 1,543 square miles, Tlaxcala is the smallest Mexican state, but it's still full of excitement and mystery. Pyramids grace the landscape, as does farmland, and the La Malinche volcano. Most of the state is on a semiarid high plateau with an average elevation of about 7,000 feet. Farmers raise barley, and corn, and ranchers raise dairy cattle, and fighting bulls. The Indigenous peoples were involved in "flower wars" with the Aztecs, in which the Aztecs attempted to capture warriors alive for sacrificial purposes, but the people of Tlaxcala kept the Aztecs at bay. And the local Nahua peoples had their own vampire.

Tlahuelpuchi

Across the world, children going through puberty have to worry about pimples, hair growth, and that awkward stage between childhood and being an adult, but in Tlaxcala, they also have to worry about becoming vampires.

Children who become Tlahuelpuchi are born to human parents, into a human family, but become a monster in adolescence. They feed by sucking blood from infants, which they have to do every month or they will perish. A Tlahuelpuchi can be male or female, but the female is always the more powerful.

They also shape-shift into the form of a vulture or a wild turkey (so many turkeys in Mexican lore).

No one can spot a Tlahuelpuchi in human form; only when they change shape can a person know what kind of monster they're dealing with. These vampires can be held at bay with onions, iron, or garlic. They can be killed by anyone, but if they are killed by a family member, the curse continues for another generation.

The Reptilian Monster of La Malinche

La Malinche is a 14,636-foot-tall volcano—dormant for at least three thousand years—popular with hikers and campers, and was once home to a monster.

This large reptile lived in a cave on the mountain and would only come down into villages to feed on the Indigenous Talax-caltec people. Its prey was usually children it would nab and drag screaming back to its lair; then it would eat them alive.

These attacks came only during the rainy season when villagers had the most difficulty chasing it through the pine forests of the mountain and onto the bare rocky peak.

According to legend, villagers eventually slew the beast, beheading it and mounting their trophy above the door of a house in the city of Puebla.

The head's no longer there, if it ever was, but I'd like to think it was.

Veracruz

A long, thin strip of a state, Veracruz has four hundred miles of beautiful Gulf coastline, but averages a width of only sixty miles; on the west side of those sixty miles is the Sierra Madre Oriental Range. Veracruz is popular with tourists, offering beaches, sport fishing, and world-class cuisine. This crescent-shaped state is graced with peaks and valleys covered with tropical rainforest. Coffee, tobacco, and vanilla are among the most grown crops; Veracruz also produces one-fourth of the country's oil. Veracruz is known as the Emerald Coast, and may, or may not be, known for the Tecolutla Monster.

Tecolutla Monster

On a night in 1969, people walking on Palmar de Susana beach discovered an enormous creature.

What became known as the Tecolutla Monster was sixty feet long, possessed a mane down its neck, had a huge beak, horns, and armor plates over its body. When the group discovered the monster, it apparently still breathed. However, afraid of what they'd found, they didn't tell anyone about it for days, but by then it was dead. By then, it also came with a smell, and at sixty feet long, I'll leave that up to you to imagine.

When the word of the creature got out, the town's mayor thought the report was of an airplane crash and sent out rescuers, according to scienceblogs.com.

When it wasn't an airplane, the rescuers turned the stinking mess over to someone who cared.

That someone involved biologists, who measured the thing. The sea serpent–shaped remains were actually sixty-six feet long, six feet wide, and weighed over twenty-four tons. Although it did have armor plates, as the discovering party reported, it was also covered in wool.

Trouble was, none of the biologists who examined the blob could identify it.

Then the theories came out. Could it be a prehistoric creature that thawed from a rogue iceberg? Maybe it was a living fossil washed up on the beach? Or, how about a dead whale?

Over the years, scientists have determined it's a sei whale, long, tubular, and beaked. Pooh.

As I've said before, scientists take all the fun out of everything.

Gulf of Mexico Coelacanth

Until Christmas 1938, the scientific world had every right to believe the lobe-finned fish, the coelacanth, had been extinct for sixty-six million years. It was a fossil, after all.

Then, right before Christmas, a fishing boat at the mouth of South Africa's Chalumna River made a strange catch. The captain of the vessel, Hendrick Goosen, had no idea what kind of fish it was, but he knew it was odd. Odd enough that he contacted South African naturalist Marjorie Courtenay-Latimer, who eventually figured out what the fish was—a living remnant from the time of the dinosaurs.

Since then, coelacanths have also been discovered off the African coasts of Tanzania, Kenya, Mozambique, Madagascar, and Comoros. A separate species of the fish has also been found in Indonesia. The African species, *L. chalumnae*, is critically endangered with fewer than five hundred specimens living. The Indonesian *L. menadoensis* is better off, with around ten thousand surviving fish.

Those are the only two species of the living fossil on the planet—wait, wait, wait.

What about the Gulf coelacanth?

In 1949, a scientist received fish scales from a strange fish caught in the Gulf of Mexico, and it was, strangely enough, from a coelacanth, and unknown to science, but known to Gulf fishermen since the seventeenth century.

That's not the only evidence to a coelacanth population living (at least at one time) in the Gulf of Mexico.

GULF OF MEXICO
COELACANTH

In 1991, cryptozoologist Gary Mangiacopra revealed the Carnegie Museum of Natural History in Pittsburgh, Pennsylvania, had on display a silver goblet with a coelacanth carved into it, according to a 1991 article in the *New York Times*. The goblet was labeled as a seventeenth century Spanish artifact, but its origin is unknown. It's not hard to imagine the goblet had reached Spain from Mexico (because it's not like the Spanish ever took treasure from the Natives, or anything).

Although there's no scientific evidence of a population of the living fossil in the Gulf of Mexico, there's a chance. Back to the coelacanth scale pulled from the Gulf. Where is it? It's lost, which is how these things go.

Yucatán

Yucatán is...well, look at a map, it's on the Yucatán Peninsula. The world-famous Maya ruins at Chichén Itzá are there, so are gorgeous colonial churches, and natural beauty, such as semiarid hills, miles of jungle, and water-filled sinkholes that dot the landscape. While tourism is important for the economy, so are fruits, livestock, and assembly plants. And there's the Cave Cow.

Wait. What? Cave Cow?

The Cave Cow

Described as a large quadruped with thick black fur, a white mane, and three-clawed "hands," the Cave Cow is certainly not cowlike.

In the 1800s, a French expedition into the jungles of the Yucatán stumbled upon the tracks of a large creature. When they followed those tracks, the beast that made them mauled a member of the party. They didn't see the creature; the party member had gone into a grove of trees, only to begin screaming in agony and

CAVE COW

fear. Before the man died, he said a monster attacked him before disappearing into the bush.

The expedition followed the trail the creature left, which included large three-clawed prints, one claw resembling the placement of a human thumb. They eventually lost the creature's trail, but one thing was certain—a Cave Cow had killed their friend.

A later British expedition witnessed a Cave Cow trudging through a swamp. Black fur, white mane, three-clawed hands.

The beast has been seen multiple times over the years, cryptozoologists postulating it may be a surviving population of giant ground sloth that inhabited North and South America from the Pliocene to the Pleistocene epoch. *Megatherium americanum* was twelve to twenty feet tall, shaggy, and possessed three-toed clawed feet, two claws in the position of human fingers, the third the thumb.

Dzulúm—Part 2

Although we covered the Dzulúm in the state of Chiapas, it was a huge white jaguar with a horse's mane that lures women into the wilderness never to be seen again, or, if the captive woman makes a deal with the creature, it takes their soul.

The Dzulúm of Yucatán is quite different. In this state, the Dzulúm appears as a man with a long flowing beard who wanders the countryside, protecting crops.

Could the Dzulúm be more different? From a terrifying, woman-nabbing jaguar to a kind custodian of nature, and savior of farmers. I'm giving the Yucatán Dzulúm ten points for style.

Zacatecas

The El Edén mine, in the Cerro de la Bufa, was one of the richest mines in Mexico, pulling up gold, silver, copper, zin, and iron from 1586 to the 1960s; however, early on, Indigenous people, enslaved by the Spanish, worked and died in that mine. Now, it's a tourist attraction. In the north-central part of the country, most of the state is in the Mexican Plateau, the Sierra Madre Occidental Mountain Range in the south, and Sierra Madre Oriental in the north. The southern bits of the Chihuahuan Desert dip into the northernmost part of Zacatecas. All that wilderness makes the perfect spot to meet Space Aliens.

Space Aliens

In 2021, an anonymous fan of the *Espooky Tales* podcast related to the hosts a tale of a 1994 space alien encounter in Mexico.

Near the unincorporated town of La Tinaja, a tiny speck on the map near the Zacatecas border with Jalisco, two young men rode home horseback from a dance when an airplane flew overhead. A strange light then appeared over the airplane, freezing the craft midair. Frightened, the men split up and headed to their own homes for the night.

That wasn't the end of the encounter.

As one of the young men put his horse in the barn, a bright light flooded his family's farm. He ran inside the house to find his family surprised he was frightened. He explains what happened, but his family was skeptical.

At bedtime, a thump came from the roof, followed by footfalls. One family member pointed out the feet dragging across the roof sounded like scraping claws. The family grabbed their guns and waited for the worst.

By morning when the worst didn't come, they told their story in town and discovered other local residents have seen the strange lights in the sky too, although they've been spared an encounter with the horrendous creature on the roof.

LA LLORONA

La Llorona—Part 2

Like the people of Costa Rica and across Latin America, Mexico also has the legend of la Llorona, the Weeping Woman. Zacatecas has its own version.

A mother visits a fortune-teller who informs her she will soon die. Worse, the fortune-teller says her children will soon die as well. Skeptical, and more than a little frightened, she returns home, kisses her children, and they all go to bed.

In the middle of the night, an enormous storm hits, filling the nearby river and pushing the water over its banks. The flood washes through her village, sweeping her house away. The mother is saved and sets out to find the home with her missing children, but they are gone.

The mother doesn't give up, walking up and down the waterways of Zacatecas, weeping as she searches for her lost babies. However, she dies before she finds them.

She returns night after night, dressed in white, crying and wailing while walking the waterways, looking for the lost children she will never find.

There are many versions of this story. Some from Spain, some from Aztec legend. They differ in the details (in most stories, she drowns her children to be with the man she loves), but the story arc of this miserable mother remains the same—she is tortured for eternity by the loss of her children to the water.

CONCLUSION

I HOPE YOU ENJOYED reading *Chasing North American Monsters* as much as I enjoyed researching and writing it. There's something about the unknown that appeals to the curious. You know, you and me.

None of us are satisfied by the explanation that the seven-foot-tall bipedal dog that whispered our name from the shadows was the neighbor's Jack Russell terrier Scraps, that the giant bat-winged beast that flew in silhouette against the moon wasn't a pterodactyl, or, in UFO (UAP, whatever) circles, the lame excuses. It was swamp gas. It was a weather balloon. It was the planet Venus. No way, man. I saw a werewolf in my yard, Jurassic Park is real, and Venus doesn't shoot out of sight in seconds at a forty-five-degree angle.

If you've doubted the existence of cryptids before now, admit that's only because you haven't seen one. Honestly, you'll feel better because that admission will open your mind to experiences and possibilities you've never dreamed of.

That's what the human experience should be about anyway. Enjoy a pint, stop at a seedy joint (not a restaurant, a joint), and try food you've never tried; love your neighbor, and be on the lookout for monsters.

They're out there. If you don't know that by the time you get to this page, I did my job wrong.

Acknowledgments

I WROTE MY FIRST monster book, *Chasing American Monsters*, in 2018; it was published in 2019, and has done quite well. So, in 2023, my publisher, Llewellyn, asked if I'd be interested in writing a follow-up.

Hmm. Lemme see. My publisher approached *me* to write a book (it usually works the other way around) about monsters (a lifelong passion), and I'd have a year to do it. (A person can do a lot of things in a year: change careers, lose fifty to one hundred pounds, grow a cavity in a tooth, watch the Lord of the Rings trilogy, maybe even twice!)

So, yeah. Sweet. No problem.

But I didn't do it by myself. No way. There are plenty of folks I need to thank who helped me along this journey. First, my family, who let me work interrupted in my basement (dungeon) office for a few hours every day to do what I do. Without their support, I would have missed deadlines, and, as an old newspaper guy, I *hate* missing deadlines. Second, Benjamin Grundy and Aaron Wright at *Mysterious Universe*, who gave me a shot to write creepy stuff for them more than a few years ago. Third, cryptozoologist Loren Coleman. He isn't only the inspiration for an entire generation of monster lovers, he's a pretty nice guy as well. You read Loren's foreword earlier in the book, and, if you're one of those people who don't read forewords, shame on you. Go back and read it, right now. It's okay, I'll wait.

Are you back? Great. You feel better, don't you?

I'd also like to thank Lee Meador, a kind, gentle old newspaperman who, during my first journalism job, taught me to tell the story by writing about what's important; my last newspaper editor Jeff Fox who taught me how to trim the fat from my writing; and the writings of the brilliant Rod Serling, who taught me as a young child as I watched *The Twilight Zone* reruns (at probably much too young an age) that monsters—the beasts that really haunt our lives—are ourselves.

I know I've missed a ton of people (if you find them, weigh them. I can guarantee you I haven't thanked an actual ton of people who assisted me with this book in some fashion), but the people I mentioned are a damn good start. Thank you, thank you, thank you.

And lastly, but truly the real first I should have thanked, is you, the reader. Without you, no book would exist.

—Jason Offutt
Maryville, Missouri
June 20, 2024

Bibliography

Introduction

Benedict, Adam. "Cryptid Breakdown: Anatomy of a Sasquatch." *Pine Barrens Institute*, August 18, 2018.

"Fouke Monster." Encyclopedia of Arkansas. Accessed February 20, 2025. https://encyclopediaofarkansas.net/entries/fouke-monster-2212/.

"Iowa Monsters: Lockridge Monster." K923. Accessed February 20, 2025. https://k923.fm/iowa-monsters-lockridge-monster/.

Long, William R. "In Brazil, U.S. Scientist Thinks He's Close to Finding Huge Sloth Thought to Be Extinct." *Los Angeles Times*, March 30, 1994.

Marietta, Nancy. "Honey Swamp Monster and Harlan Ford's Film." *Nexus Newsfeed*, September 27, 2018.

Salter, Jim. "40 Years Later, Debate Over 'Momo' Lingers." *Columbia Missourian*, July 14, 2012.

Wierima, Brian. "The Hunt for the Elusive Bigfoot." *Detroit Lakes Tribune*, November 17, 2011.

Belize

Gonzalez, Mary. "Belizean Folktales: El Cadejo." October 12, 2018.

Penafiel, Carolina Bucheli. *El Duende: Folktale, Oral History, and the Construction of Gendered and Racialized Discourses in Quito*. University of New Mexico, 2020.

Staff writer. "The Legend of 'La Xtabay.'" *The Yucatán Times*, November 1, 2019.

Canada

Alberta

Azpiri, Jon, and Aaron McArthur. "B.C. Bigfoot Lawsuit a Big Waste of Time, Critics Say." *Global News*, August 14, 2018.

Lapointe, Chris. "The Haunting of the Lakeland: The Legend of the Fort Kent Wendigo." *Lakeland Connect*, October 30, 2020.

Mitchell, Laine. "A Massive Lake in Alberta Is Said to Be Home to a Monster Named 'Kinosoo.'" *Daily Hive Canada*, March 14, 2024.

Paranormal Rona. "Counting the Cryptids: *Edmonton Examiner*." Paranormal Explorers, June 13, 2016. https://paranormalexplorers.com/2016/06/13/counting-the-cryptids-edmonton-examiner/.

Perry, Douglas. "Bigfoot Hunters Sue State of California for Denying Existence of Sasquatch." *The Oregonian*, February 14, 2018.

Schmunk, Rhianna. "Sasquatch Tracker's Lawsuit Tossed by B.C. Supreme Court." CBC, September 5, 2018.

British Columbia

Anderson, David C. "It's Hard to Prove That Something, Even a Monster, Doesn't Exist." *The New York Times*, January 20, 1974.

Jones, Robert. "Local Legends: Devil Monkeys & Gugwes." *Caledonian Record*, November 24, 2023.

Mackie, John. "This Week in History, 1937: The Saga of Caddy Cadborosaurus, Vancouver Island's Own Sea Monster." *Vancouver Sun*, February 9, 2024.

Murray, Nick. "B.C. Group on the Hunt for Cadboro Bay Sea Monster." *Victoria News*, May 16, 2019.

Radford, Benjamin. "Cammy: A New Canadian Lake Monster?" *Live Science*, September 21, 2009.

Rockingham, Graham. "Relative of Loch Ness Monster: Nessie's Canadian Cousin Warms to Tourists." *Los Angeles Times*, October 26, 1986.

Staff reporter. "Scientist Seeks Mythical Cadborosaurus." CBC, 1999. 6:58. https://www.cbc.ca/player/play/video/1.3594202.

Trudeau, Scott. "Ogopogo Still Swims in Our Memories." *Penticton Herald*, July 14, 2017.

Walbran, Captain John T. "What Lurks at the Bottom of B.C.'s Cameron Lake?" *CTV News*, October 4, 2009.

Manitoba

Bernhardt, Darren. "Keep Your Camera Handy: Stories of Manitoba Lake Monsters Told for Centuries but Proof Remains Elusive." CBC, June 3, 2018.

Cain, Patrick. "As Climate Warms, Grizzly Bears and Polar Bears Interbreed." *Global News*, May 25, 2016.

Lunney, Doug. "Manipogo Park, Magnetic Hill, Bigfoot Prints & Other Unusual Manitoba Attractions." *Winnipeg Sun*, August 10, 2013.

Staff writer. "'Bigfoot' Tape Thrills Northern Community." CBC *News*, April 21, 2005.

New Brunswick

Cox, William T. *Fearsome Creatures of the Lumberwoods*. Press of Judd & Detailer, 1910.

Slaten, Michael. "Eastern Cougar Subspecies Declared Extinct." *The Daily Cougar*, January 26, 2018.

Wright, Bruce. "New Brunswick." May 29, 2018. https://www.soulask.com/new-brunswick/.

Wright, Julia. "How a Sea Monster Myth Was Born in West Saint John." CBC, January 28, 2020.

Newfoundland and Labrador

Adey, Jane. "What Lurks in Crescent Lake? Meet Cressie, N.L.'s Water Monster." CBC, November 2, 2019.

Crantz, David. *History of Greenland*. Brethren's Society for the Furtherance of the Gospel Among the Heathen, 1773.

Jarvis, Dale. *Wonderful Strange: Ghosts, Fairies, and Fabulous Beasties*. Flanker Press, 2005.

McMillan, Kelly. "Cryptids of Canada the True North Strong and... Weird." *Misfits and Mysteries*, May 28, 2021.

Rieti, Dr. Barbara. *Strange Terrain: The Fairy World in Newfoundland*. ISER Books, 1991.

Rose, Sydney. "The Giant Squid of Thimble Tickle." *Atlas Obscura*, March 11, 2021.

Whiffen, Glen. "Bonavista Newfoundland Sea Creature 2000." *The Telegram*, April 6, 2000.

Nova Scotia

Ashton, John. "Sea Monsters of the Strait." *Saltwire Network*, July 27, 2014.

Hebda, Andrew. *The Serpent Chronologies: Sea Serpents and Other Marine Creatures from Nova Scotia's History*. Nova Scotia Museum Publications, 2015.

Staff, PNI Atlantic. "Was It a Sasquatch on Green Hill?" *Saltwire Network*, January 10, 2016.

Staff writer. "Sea Monster Sightings off Nova Scotia Documented in Free E-Book." CBC, November 11, 2015.

Wallis, Wilson D., and Ruth Sawtell Wallis. *The Micmac Indians of Eastern Canada*. The University of Minnesota Press, 1955.

Ontario

Bonaparte, Jordan. "Nighttime Podcast Recap: Was It a Sasquatch in Pictou County, N.S.?" *Global News*, March 14, 2019.

Brody, Erin. "Jack and Joseph Fiddler: Wendigo Hunters." *Siren Media Group*, May 26, 2021.

Godfrey, Linda. "New Dogman Reports in Ontario and New Jersey." Lindagodfrey's Blog, December 21, 2015. https://lindagodfrey.com/2015/12/21/new-dogman-reports-in-ontario-and-new-jersey/.

Goldstein, Lorrie. "Tunnel Monster of Cabbagetown?" *Toronto Sun*, March 25, 1979.

Mahoney, Jill. "Big Trout Lake 'Monster' Sparks Internet Debate." *The Globe and Mail*, May 21, 2010.

Nickell, Joe. "Investigators Search for Canadian Lake Monster." *Live Science*, October 14, 2005.

Ostberg, René. "Wendigo." *Britannica*, 2023.

Smythe, Eleanore. "Ontario, Canada Encounter." Dogman Encounters. Accessed February 21, 2025. https://dogmanencounters.com/ontario-canada-encounter/.

Staff writer. *Kingston Gazette and Religious Advocate.* August 14, 1829.

Staff writer. "The Legend of Old Yellow Top." Bigfoot Encounters. Accessed November 22, 2018.

Prince Edward Island

Peters, Hammerson. "Legends of Prince Edward Island." *Mysteries of Canada.* September 1, 2022.

Staff writer. "West Point Lighthouse Stories & Folklore." Accessed February 21, 2025. https://peilighthousesociety.ca/index.php/history/lighthouses/west-point.

Statistics Canada. "Special Interest Profile, 2021 Census of Population: Profile of Interest: Ethnic or Cultural Origin." National Household Survey, 2021.

Watson, Julie V. *Ghost Stories and Legends of Prince Edward Island.* Dundurn Press, 2018.

Wiley, Nathan. "Prince Edward Island Bigfoot Revisited." *The Endangered Left*, September 8, 2015, 1:42. www.youtube.com/watch?v=Mj94gUtCblY.

Quebec

Beacon, Bill. "The Lake Champlain Monster? How About Pohenegamook, Quebec's Loch Ness?" *UPI*, May 21, 1982.

Boisvert, Jacques. "Canada's Lake Creature: Memphré." *Welcome to Ogopogo Country*, 2001.

Cucco, Kyle. "The Werewolf of Quebec." This Is Canada, November 8, 2022.

Gagnon, Claude, and Michel Meurger. *The Monsters of Quebec's Lakes*. Stanké Publisher, 1982.

Langlois, Hubert. "Beware the 'Loup-Garou.'" CBC Archives: Quebec Now, December 25, 1973.

Staff writer. "Bigfoot Sighting Reported by Cree Hunter Near Wemindji, Que." CBC, August 29, 2013.

Staff writer. "Creepy Creature or Window Smudge? Video Sparks Online Debate." KHQ, August 4, 2018.

Staff writer. "'Memphre': The Lake Memphremagog Monster." Charles Fort Institute, August 8, 2011. https://forums.forteana.org/index.php?threads/memphre-the-lake-memphremagog-monster.9236/.

Staff writer. "The Mysterious History of Champ." Lake Champlain Region. Accessed February 21, 2025. https://www.lakechamplainregion.com/heritage/champ.

Staff writer. "Native American Legends: Chenoo (Chenu)." Native Languages. Accessed November 22, 2020.

Saskatchewan

Archer, Dan. "Cryptozoology: A Look at Mysterious Creatures." *SaskToday*, September 14, 2018.

Brown, Emily. "Terrifying Footage Shows Huge 7ft Wolf Chasing a Dog in the Woods." *LADbible*, June 19, 2018.

Casey, Liam. "'Large-Bodied' Canadian Wolves to Help Keep U.S. Moose Population in Check." *National Post*, January 30, 2019.

Johnson, Will. "Chasing the Deep Bay Monster." *Nelson Star*, August 18, 2015.

Maxwell, Nigel. "Fishing for the Turtle Lake Monster." *Prince Albert NOW*, June 30, 2013.

Staff writer. "Mystery Beast Blamed for Killing 3 Dogs." CBC, August 10, 2006.

Northwest Territories

Bernhardt, Darren. "Did Giants Roam Canada's Northwest Territories—or Do They Still?" CBC, September 30, 2017. https://www.cbc.ca/news/canada/north/giants-lakes-footprint-mythology-1.4309431.

Bird, Hilary. "N.W.T. Man Tells of Encounter with Nàhgą—the Tlicho Sasquatch—Following Boat Accident." CBC, July 28, 2016. https://www.cbc.ca/news/canada/north/whati-man-nahga-bushmen-encounter-1.3698240.

Campbell, Daniel. "The Hero of the Dene." *Up Here Magazine*. Accessed April 8, 2024. https://www.uphere.ca/articles/hero-dene.

Hess, Brooke. "Secrets of the Nahanni: The Valley of Headless Men." *The Outdoor Journal*, September 12, 2018. https://www.outdoorjournal.com/secrets-nahanni-valley-headless-men/.

Nunavut

"Bigfoot on Hudson Bay." *Nunatsiaq News*, June 29, 2001. https://nunatsiaq.com/stories/article/bigfoot_on_hudson_bay/.

Boas, Franz. "The Central Eskimo." In *Sixth Annual Report of the Bureau of Ethnology, 1884–1885*. 399–669. Bureau of American Ethnology, 1889.

Eberhart, George M. *Mysterious Creatures: A Guide to Cryptozoology*. CFZ Press, 2002.

"Ijirait." Inuit Myths & Legends. Accessed December 6, 2021. http://www.inuitmyths.com/ijirait.htm.

"Mahaha." Inuit Myths & Legends. Accessed December 6, 2021. https://www.inuitmyths.com/mahaha.htm.

Toombs, Terrye. "Alaska Folklore: Five Mythical Creatures of the Last Frontier." *Anchorage (Alaska) Daily News*, June 12, 2012.

Yukon

Benedict, Adam. "Cryptid Profile: Saytoechin (AKA: Beaver-Eater)." *Pine Barrens Institute*, August 18, 2018.

Dupuy, Georges. "The Monster of 'Partridge Creek.'" *Je sais tout* and *The Strand Magazine*, 1908.

Roden, Barbara. "Golden Country: Past, Present, and Beyond: Camels in the Cariboo." *Ashcroft-Cache Creek Journal*, September 6, 2016.

Staff writer. "Native American Legends: Wechuge." Native Languages. Accessed December 5, 2021.

Staff writer. "Sasquatch Sighting Reported in Yukon." CBC, July 13, 2005.

The Caribbean

Andros

Cáceres, Melissa. "On the Hunt for the Sea Monster Eating Swimmers in Paradise." *New York Post*, May 5, 2016.

NEA contributor. "Chickcharney, Caribbean Folklore." *Northend Agents*, October 12, 2021.

Antigua and Barbuda

Narine, Aminta Kilawan. "Jumbee Time!" *The West Indian*, October 30, 2018.

Staff writer. "Choorile Jumbie (Jumbee)." *Village Voice*, August 14, 2021.

Telford, Moss. "Jumbie Story." *Stabroek News*, July 2, 2016.

Bay Islands

Haines, Lindsay. "Obeah Is a Fact of Life, and Afterlife, in the Caribbean." *The New York Times*, September 10, 1972.

Staff writer. "The Legend of La Siguanaba." *Espooky Tales*, December 2, 2020.

Cuba

Ginzo, Dr. Alberto Alvarez. *Caleidoscopio*. CreateSpace, 2018.

Grenada

Blakeslee, Vanessa. "Anancy Today: The Role of Folklore in the Afro-Caribbean Aesthetic." *Medium Magazine*, February 3, 2020.

Compton, Fiona. "The Story of the Soucouyant and the Loogaroo." *Know Your Caribbean Podcast*, October 31, 2023.

Malm, Sara. "Is It Fish or Just Foul? 'Mutant' Sea Creature with a Nose, Feet, Tail and WINGS Baffles Caribbean Island." *UK Daily Mail*, February 1, 2016.

"Mamadjo." *The Caribbean Dictionary*. Accessed February 22, 2025. https://wiwords.com/words/mamadjo.

Polk, Patrick. "African Religion and Christianity in Grenada." *Caribbean Quarterly* 39, no. 3/4 (September/December 1993).

Hispaniola

Cummings, Monique-Marie. "Elizabeth Acevedo Sees Fantastical Beasts Everywhere." *Smithsonian Magazine*, May 5, 2020.

Hernandez, J. A. "El Coco, El Cucuy: The Child Eater." *Into Horror History*, March 28, 2023.

Jailler, Fayida. "Clairvius Narcisse: The Alleged Real-Life Haitian Zombie." *Travel Noire*, October 22, 2021.

Staff writer. "4 Dominican Urban Legends of Dark Creatures That Go Bump in the Night." Pisqueya. Accessed March 1, 2024.

Staff writer. "Pontarof." A Book of Creatures. Accessed June 11, 2021. https://abookofcreatures.com/2021/06/11/pontarof/.

Zuckerman, Edward. "The Natural Life of Zombies." *Outside Magazine*, October 15, 2013.

Jamaica

MacEdward, Leach. "Jamaican Duppy Lore." *The Journal of American Folklore* 74, no. 293 (July/September, 1961).

Puerto Rico

Jarry, Johnathan. "The Mythical Creature Known as the Chupacabra Walked out of a Movie." *McGill University, Office for Science and Society: Separating Sense from Nonsense*, June 30, 2023.

Pethick, Kris. "Here's Why the Coquí Frog Is the Symbol of Puerto Rico." *Culture Trip*, November 29, 2024.

Plá, Lucy. "Los Chupacabras: The Interview." *UFO Digest*, March 20, 1996.

St. Lucia

Staff writer. "Papa Bois." *Caribbean Reads.* Accessed October 12, 2023.

Trinidad and Tobago

Bissessarsingh, Angelo. "Douens and Other Folklore." *Trinidad and Tobago Guardian,* June 30, 2013.

Bissessarsingh, Angelo. "Legend of the La Diablesse." *Trinidad and Tobago Guardian,* November 26, 2013.

Morris, Kirt. "Trini Folklore: Mermaids." *Trini in Xisle,* April 9, 2019.

Parsanlal, Nneka. "Fables: Fairymaids." *Trinidad and Tobago Loop News,* July 20, 2020.

Costa Rica

Donkin, R. A. "The Peccary: With Observations on the Introduction of Pigs to the New World." *Transactions of the American Philosophical Society* 75, part 5 (1985).

Mateo, Don. "Costa Rican Urban Legends: Truths and Myths Unfold." *The Tico Times,* May 9, 2024.

McCarthy, Andrew. "Costa Rican Creepy Tales." *A.M. Costa Rica,* October 26, 2021.

Staff writer. "Spine-Tingling Costa Rican Folklore Tales." *The Tico Times,* October 28, 2023.

El Salvador

Molin, Eva. "El Cipitio." *USC Digital Folklore,* May 12, 2016.

Muth, Linda. "The Legend of the Priest with No Head." Walking with El Salvador, October 28, 2021. https://blog.walkingwithelsalvador.org/2021/10/the-legend-of-priest-with-no-head.html.

Perez, Melissa. "Spooktacular Myths & Legends of El Salvador." *Life ALa Melly*, November 1, 2021.

Reyes, Aleyda. "The Legend of the Cuyancúa, Heritage of the Mayans." *Guanacos*, June 1, 2022.

Staff writer. "El Salvador." Elsalvadorfdint.weebly. Accessed January 21, 2023.

Greenland

Bach Kreutzmann, Maria, and Coco Apunnguaq Lynge. *Mythical Monsters of Greenland: A Survival Guide*. Inhabit Media, 2024.

Kristjánsdóttir, Ajaana Olsvig. "Greenlandic Myths and Legends." *Greenland Adventure Tours*, August 13, 2020.

Por, Tanny. "Greenland Myths and Legends." Visit Greenland. Accessed June 1, 2024.

Staff writer. "5 Mythical Creatures from Greenland." *Arctic Business Journal*, January 12, 2021.

Staff writer. "The Greenlandic Tupilak." Visit Greenland. Accessed June 2, 2024.

Guatemala

Aroche, Karin. "Legend of El Sombrerón in Guatemala." *Aprende Guatemala*, October 23, 2021.

Gordon, George Byron. "Guatemala Myths." In *The Museum Journal*. Penn Museum, University of Pennsylvania, September 1915.

Jones, Jonathan. "A Brief History of the Aztec Empire." *The Guardian*, September 16, 2009.

Navarro, Hector. "El Sombrerón of Guatemala." *FactsChology*, February 15, 2023.

Staff writer. "Huay Chivo, the Yucatecan Legend." *The Yucatán Times*, April 25, 2023.

Staff writer. "Ix-hunpedzkin." A Book of Creatures. Accessed May 12, 2017. https://abookofcreatures.com/2017/05/12/ix-hunpedzkin/.

Staff writer. "The Legend of the Huay Chivo." *Espooky Tales*, December 24, 2020.

Staff writer. "Quetzalcoatl." A Book of Creatures. Accessed June 22, 2020. https://abookofcreatures.com/2020/06/22/quetzalcoatl/.

Honduras

Davis, Frederick R. "The Man Who Saved Sea Turtles: Archie Carr and the Origins of Conservation Biology." *Oxford University Press*, July 2007.

Reyes, Fernanda María Martínez. "The Taconuda That Appears to Taxi Drivers in Tegucigalpa." *Corpus Deliteratura Oral*, Jaén University, October 8, 2010.

Nicaragua

Conzemius, Eduard. "Ethnographical Survey of the Miskito and Sumu Indians of Honduras and Nicaragua." *Bureau of American Ethnology Bulletin*, 1932.

Ferenczi, Natasha-Kim. "What If There Is a Cure Somewhere in the Jungle? Seeking and Plant Medicine Becomings." *Department of Sociology and Anthropology Faculty of Arts and Social Sciences*, Simon Fraser University, December 10, 2018.

Kiprop, Joseph. "Do the Minhocão Really Exist?" *World Atlas*, May 23, 2018.

Leiva, Katherine. "La Carreta Nagua." *Life of Leiva*, October 31, 2023.

Lopez, T. "La Carreta Nagua." *USC Digital Folklore Archives*, May 7, 2018.

Staff writer. *Encyclopaedia of Cryptozoology*. Accessed February 22, 2025. https://cryptidarchives.fandom.com/wiki/Glyptodont.

Staff writer. "Minhocão." *Encyclopedia of Cryptozoology*. Accessed May 12, 2024. https://cryptidarchives.fandom.com/wiki/Minhocão.

Verne, Jules. "Eight Hundred Leagues on the Amazon." *The Steam House*, 1881.

Panama

Aroeste, Alex. "La Tulivieja—A Panamanian Demon/Monster." *USC Digital Folklore Archives*, September 26, 2016.

Delgado, Maggi. "Brujas of Yesterday, Their Legacy Today." *The Graduate Center*, City University of New York, June 2022.

Gonzalez, Geraldine. "Halloween in Panama: Popular Myths and Legends." *Live and Invest Overseas*, October 29, 2021.

Sherwood, Joseph. "The Legend of La Tulevieja: The Woman Wandering Panama's and Costa Rica's Rivers." *A Little Bit Human*, April 15, 2022.

Staff writer. "Malpelo Monster." Cryptidz Fandom. Accessed February 22, 2025. https://cryptidz.fandom.com/wiki/Malpelo_Monster.

Staff writer. "Mystery Beast Beaten to Death by Kids in Panama." *New York Post*, September 17, 2009.

Staff writer. "Photo: Mystery 'Alien-Beast' in Panama Is Likely a Sloth." *Mongabay News*, September 19, 2009.

Trejos, Edilberto González. "Panamanian Myths and Legends." *The Melting Pot Magazine*, September 9, 2007.

The United States of America

Alabama

Deal, Marissa. "Port City Legends." *Mobile Bay Magazine*, October 31, 2023.

Helean, Jack. "13 Years Ago NBC 15 Aired the Crichton Leprechaun Story, the Rest Is Internet History." WPMI, March 17, 2019.

Hill, Emily. "Crichton Leprechaun Revisited: Meet the Man Who Discovered the Legend." *The Mobile Real-Time News*, March 9, 2015.

Kazek, Kelly. "Original Drawing of Legendary Wolf Woman of Mobile Discovered in Archives." AL. Accessed February 22, 2025. https://www.al.com/living/2015/10/original_drawing_of_legendary.html.

Whitehead, Vera. "The Downey Booger." The Free State of Winston. Accessed May 29, 2024. http://www.freestateofwinston.org/downeybooger.htm.

Alaska

Mackal, Roy P. *Searching for Hidden Animals*. Doubleday, 1980.

Arizona

Herreras, Mari. "After Nine-Hour Desert Bike Ride, a Guy Sees a Lizard Man." *Tucson Weekly*, February 17, 2014.

Arkansas

Jameson, W. C. *Ozark Tales of Ghosts, Spirits, Hauntings, and Monsters*. Progreso Publishing Group, 2015.

Ogilvie, Craig. "Legendary Arkansas Monsters Have Deep Roots in History." *Arkansas Tourism*, October 8, 2002.

Staff writer. "Other 'Creatures' Skulk About State." *Arkansas Democrat Gazette*, November 16, 2014.

California

Kohlruss, Carmen. "'Cute Cryptid.' Fresno Nightcrawler Is a Paranormal Darling with Supernatural Fandom." *The Fresno Bee*, October 27, 2021.

McGlone, Ron. "Strange Creature Reported in Carmel Area." *The Highland County Press*, December 19, 2014.

Staff writer. "A Monster Lurks in Lake Elsinore." *The Los Angeles Times*, September 13, 1934.

Staff writer. "Letter to the Editor: Lone Pine Mountain Devil?" *Sierra Wave Media*, November 12, 2013.

Colorado

Alexander, J. H. "The Legendary North American Lake Creature! Blue Dilly, the Blue-Dillon Monster." Accessed April 12, 2024. http://www.bluedillonmonster.com.

Benjamin, Shane. "Bigfoot Sighting Near Silverton Goes Viral." *The Durango Herald*, October 12, 2023.

Mitchell, Alex. "Bigfoot 'Spotted' in Colorado in Broad Daylight—and It's All on Camera: 'We're Convinced.'" *The New York Post*, October 11, 2023.

Staff writer. "Dillon Reservoir: Play on and Around Denver Water's Largest Reservoir." Denver Water. Accessed April 12, 2024. https://www.denverwater.org/recreation/dillon-resevoir.

Connecticut

Bard, Megan. "In 1854, Vampire Panic Struck Connecticut Town." *The Register Citizen*, November 2, 2008.

Brink, Susan. "Before COVID, TB Was the World's Worst Pathogen. It's Still a 'Monster' Killer." NPR, February 13, 2022.

Staff writer. "Reported Tuberculosis in the United States, 2023." Centers for Disease Control and Prevention, October 18, 2023.

Delaware

Beaumont, Wade. "Discovering Delaware's Daunting Cryptids." Hangar 1 Publishing. Accessed June 19, 2024. https://hangar1publishing.com/blogs/cryptids/delaware-cryptids.

Florida

Staff writer. "Lake Tarpon Monster." Cryptid Wiki. Accessed May 13, 2024.

Georgia

Phillips, Y. "Raptor Sighting in Georgia." Cryptid Wiki. Accessed June 23, 2023.

Hawaii

Staff writer. "Kamapua'a—The Pig God." Private Tours Hawaii, July 26, 2015.

Idaho

Patrick, Sholeh. "Meet Your N. Idaho Lake Monster, Paddler." *Coeur d'Alene Post Falls Press*, September 17, 2019.

Illinois

Lutz, Eric. "Behold the Persistent Legend of Our Deep Blue Waters: The Lake Michigan Sea Serpent." *The Chicago Tribune*, August 7, 2020.

Indiana

Staff writer. "Mysterious Indiana—Aggressive Monsters and Big Serpents." *The Weekly Rambler*, June 15, 2020.

Iowa

Clough, Eliot. "Terror in Our Own Backyard: Four Monsters in Iowa." 97.7KCRR, October 28, 2022.

Kansas

Dewey, Ernest. "Monster Turns Out to Be a Plain Old Foopengerkle." *The Salina Journal*, November 23, 1952.

Gilliland, Steve. "Exploring Outdoors Kansas: The Legend of Sinkhole Sam." *The Hays Post*, September 10, 2023.

Kentucky

White, Douglas. "Spottsville Monster May Be Subject of Documentary." *The Gleaner*, August 15, 2018.

The Travel Channel. "Mantis Man, Spottsville Monster, Tornado Phantoms." *Monsters and Mysteries in America*, season 3, episode 8. Aired March 18, 2015.

Louisiana

Jones, Terry L. "The Wild Girl of Catahoula." *Country Roads Magazine*, March 15, 2017.

Taylor, Trish Cook. "The Wild Girl of Catahoula." *Jena Times-Olla-Tullos Signal*, October 25, 2022.

Maine

Bavoso, Katharine. "Maine Mysteries: Cassie the Casco Bay Sea Serpent." *News Center Maine*, October 26, 2016.

Steinauer-Scudder, Chelsea. "The Great Sea Serpent of Casco Bay." *Emergence Magazine*, June 16, 2018.

Maryland

Staff writer. "Mysterious 'Dwayyo' on Loose in County: Don't Mess with It." *The Frederick News-Post*, November 27, 1965.

Massachusetts

D'Agostino, Thomas. "Creatures of the Bridgewater Triangle." *The Yankee Express*, May 10, 2022.

Michigan

Staff writer. "Native American Legends: Flying Head (Big Heads)." Native Languages. Accessed June 2, 2024.

Minnesota

Ross, Carly. "Have You Ever Heard of the Minnesota Dog-Man?" KDHL 920AM 97.9FM, August 6, 2020.

Mississippi

Staff writer. "Case File #5: The Chatawa Monster: Mississippi's Bigfoot." Hinds Community College, September 7, 2023.

Missouri

Randolph, Vance. *We Always Lie to Strangers: Tall Tales from the Ozarks*. Oxford U.P., 1951.

Staff writer. "Jimplecute." Accessed June 22, 2024. https://creatures-of-myth.fandom.com/wiki/Jimplecute.

Montana

Basner, Dave. "'Naked Alien' Photographed in Part of Montana Known for UFO Sightings." WBZ NewsRadio 1030 AM, January 5, 2022.

Nebraska

Trudell, Tim. "Capturing the "Oakland Creature": Four Decades Ago, Confusing Sightings and a Mass Theory Terrified a Small Nebraska Town." *Flatwater Free Press*, March 3, 2023.

Nevada

Fiorentino, Mark. "Curse of the Pyramid Lake Water Babies." *The Granby Drummer*, August 31, 2019.

Staff writer. "Water Babies of Pyramid Lake—Legend." USC Digital Folklore Archives, March 30, 2024.

New Hampshire

Citro, Joseph A. *Weird New England: Your Guide to New England's Local Legends and Best Kept Secrets.* Union Square & Co., 2010.

Staff writer. "Tales of Old Derry: The Legend of the 'Derry Fairy.'" *Derry News*, November 13, 2008.

New Jersey

Izzo, Michael. "Lake Hopatcong's Original Sea Creature." *Morris County Daily Record*, July 19, 2014.

Staff writer. "Record Fish Caught in Lk. Hopatcong, in Beginning of Fruitful Fishing Season." *Bergen Record*, April 15 2016.

New Mexico

Torres, Larry. *Los Cocos y Los Coconas: Bogey Creatures of the Hispanic Southwest.* New Mexico State University, 1995.

New York

Israel, Steve. "Two-Headed Trout Lives on in Fish Tale." *Times Herald-Record*, April 1, 2000.

Pitcher, Glenn. "Have You Seen the Legendary Two-Headed Trout of Roscoe, New York?" 98.1 The Hawk, April 4, 2022.

Taylor, Traci. "Unveiling Five Mysterious Cryptids of New York." 98.1 The Hawk, April 10, 2024.

North Carolina

George, Dustin. "TV Crews Come to Casar in Search of Knobby." *Shelby Star*, July 1, 2020.

North Dakota

Schwarz, Donna. "Mythical Creature Tradition Continues." *McLean County Independent*, June 24, 2020.

Staff writer. "Monsters on the Plains." *High Plains Reader*, October 25, 2017.

Ohio

Wiley, Chelsea. "The 11 Most Bizarre Cryptids and Monsters from Ohio." *Columbus Navigator*, October 30, 2023.

Oklahoma

Staff writer. "The Legend of Oklahoma's Most Terrifying Monster Will Send Chills Down Your Spine." *Only in Your State*, September 18, 2023.

Staff writer. "Native American Legends: Stikini." Native Language. Accessed June 17, 2024.

Oregon

Eberhart, George M. *Mysterious Creatures: A Guide to Cryptozoology*. CFZ Press, 2015.

Girgis, Lauren. "A Field Guide to Washington's Mysterious, Monstrous Cryptids." *The Seattle Times*, October 27, 2023.

Staff writer. "Amhuluk." Accessed January 21, 2019. https://abookofcreatures.com/2019/01/21/amhuluk/.

Pennsylvania

Thomhave, Kalena. "5 Legendary Creatures Believed to Live in Pennsylvania." *The Keystone*, December 20, 2022.

Rhode Island

Muise, Peter. "A Werewolf in Pawtucket, Rhode Island." *New England Folklore*, August 1, 2020.

Rosales, Albert. *Humanoid Encounters: 2000–2009*. Independently published, 2021.

South Carolina

Dodson, Braley. "10 Urban Legends in South Carolina." WBTW, August 13, 2021.

South Dakota

Benedict, Adam. "Cryptid Profile: The Thunder Horse." *The Pine Barrens Institute*, August 19, 2018.

Tennessee

Murphy, Elias. "Not-Deer, Not Going into the Woods." *East Tennessean*, January 23, 2023.

Texas

Chavez, Adriana M. "Horizon City's Monster." *El Paso Times*, July 31, 2003.

Cohen, Li. "'Mystery Animal' Caught on Camera in Texas State Park Prompts Investigation: 'The Heck Is That?'" *CBS News*, April 10, 2023.

Holt, Kari Anne. "Typewriter Rodeo: The Beast of Bear Creek." *Texas Standard*, October 13, 2023.

Long, Trish. "Horizon City Monster Sparks Fascination in Some Borderland Residents." *El Paso Times*, March 3, 2023.

Mayes, Michael. "The Beast of Bear Creek." *Texas Cryptid Hunter*, September 10, 2012.

Olive, John. "Texas Cryptids: Beasts, Monsters and a Batman." *The Corpus Christi Caller Times*, October 12, 2023.

Utah

Nelsen, Braden. "The Ghosts and Monsters of the Great Salt Lake." *The Davis Journal*, October 6, 2023.

Vermont

Hallenbeck, Brent. "Vampires in Vermont? That's What Some People Thought in 1792." *Burlington Free Press*, August 2, 2022.

Virginia

Romano, Anthony. "Virginia's Legendary Cryptids." Hangar 1 Publishing. Accessed June 19, 2024.

Washington

Dever, Jim. "Centuries-Old Unsolved Mysteries Deep Within Lake Chelan." KING5, November 5, 2021.

Roberts, Dan. "Strange Tales of the Thing That Lurks in Lake Chelan." KISSFM, October 12, 2022.

West Virginia

Reed, Daniel A. "The Curious Case of the Grafton Monster." *The Skeptical Inquirer*, August 8, 2024.

Tarley, Cavan. "Grafton Celebrates the 60th Anniversary of the First Grafton Monster Sighting." *The Dominion Post*, June 11, 2024.

Wisconsin

Lewis, Chad. *The Wisconsin Road Guide to Mysterious Creatures.* On the Road Publications, 2011.

Wyoming

Schwamle, Bill. "You Will Freak Out When You See This 'Monster' on Casper Mountain." 95.5 My Country, July 19, 2022.

The United Mexican States

Aguascalientes

Bitto, Robert. "Legends of the Chichimeca." *Mexico Unexplained*, March 21, 2021.

Staff writer. "'Giant' Recorded on Top of Hill in Aguascalientes, Mexico." *Nexus Newsfeed*, January 2, 2023.

Baja California

Bitto, Robert. "Hombre Pájaro, Mexico's Bird Man Creature." *Mexico Unexplained*, September 18, 2022.

Baja California Sur

Bitto, Robert. "The Boy with Horrible Teeth." *Mexico Unexplained*, April 18, 2021.

Staff writer. "Hotel California History." Accessed February 23, 2025. http://hotelcaliforniabaja.com/press/HC_History.html.

Campeche

Sanderson, Ivan T. *Abominable Snowmen: Legend Come to Life.* Pyramid Books, 1968.

Staff writer. "Huay Chivo, the Yucatecan Legend." *The Yucatán Times*, April 25, 2023.

Chiapas

Harris, Chris. "The Myth of the Dzulúm and Patriarchal Masculinity in Rosario Castellanos's Balún-Canán." *Bulletin of Hispanic Studies*, January 2011.

LaBrie, Laura. "The Mayan Alux: Powerful Magic in Mexico." *The Edge Magazine*, August 7, 2023.

Rupiah, Emiliano. "Werewolf of the Mexican South." *Discovery*, May 13, 2021.

Shuker, Karl P. N. *The Beasts That Hide from Man: Seeking the World's Last Undiscovered Animals*. Paraview Press, 2003.

Slater, W. C. "The 'Abominable Snowmen': Footprints in Mexico." *Times of London*, August 2, 1937.

Staff writer. "Mexican President Posts Photo of What He Claims Is a Maya Elf." *The Guardian*, February 27, 2023.

Sutherland, A. "Alux: Little Mythical Troublemaker and Guardian of Cornfields in Mayan Folklore." *Ancient Pages*, June 11, 2019.

Chihuahua

Bitto, Robert. "More Legends from the State of Chihuahua." *Mexico Unexplained*, January 14, 2024.

Chance, Ryan. "Could Abnormal Skull Be from an Alien?" *Houma Today*, February 26, 2013.

Pye, Lloyd. *The Starchild Skull—Genetic Enigma or Human-Alien Hybrid?* Bell Lap Books, 2007.

Roberts, David. "In the Land of the Long-Distance Runners." *Smithsonian Magazine*, April 30, 1998.

Staff writer. "Rarámuri Legend: The Ganoko. Were There Giants in Chihuahua?" *The Chihuahua Post*, December 19, 2023.

Coahuila

Binnall, Tim. "Purported Mothman Photo Resurfaces & Goes Viral on Social Media in Mexico." Coast to Coast AM, April 5, 2024.

Corrales, Scott. "Mexico: Four-Armed 'Grey Monster' in Monclova." *Inexplicata—The Journal of Hispanic Ufology*, April 30, 2010.

Colima

Mayes, Michael. "Can the Xoloitzcuintli Explain Texas Chupacabras Sightings?" *Texas Cryptid Hunter*, May 22, 2019.

Staff writer. "Meet the Xolo: Chupacabra Mystery Solved?" KHOU-11, July 21, 2010.

Towner, Myriah. "Collecting Samples for Your Home Planet? Dark UFO Resembling a Flying Horse Is Caught on Film Hovering by Erupting Mexican Volcano." *UK Daily Mail*, January 30, 2015.

Durango

Bitto, Robert. "The Lerdo Monster, a New Chupacabra?" *Mexico Unexplained*, December 10, 2023.

Evans, Sophie Jane. "Shocking Moment an Owl Is Interrogated by Superstitious Mexican Villagers After They Set It on Fire 'For Being a Witch.'" *UK Daily Mail*, August 7, 2014.

Staff writers. "Mapimí Silent Zone." *Atlas Obscura*, September 24, 2009.

Staff writer. "La Zona de Silencio (the Zone of Silence), and UFO Sightings." *Espooky Tales*, February 18, 2021.

Guanajuato

Bitto, Robert. "The Mysterious Craters of Valle de Santiago." *Mexico Unexplained*, November 4, 2019.

Rocha, J. L. "Legends of Love from Guanajuato." July 20, 2020. https://joseluisrocha.com.

Guerrero

Doherty, Ruth. "Mysterious 'Sea Monster' Washes Up on Beach in Mexico." *Yahoo! News*, March 10, 2016.

Pardo, Gerardo Avila. *La Trampa del Chaneque*. Universidad Veracruzana Intercultural, 2009.

Hidalgo

Bitto, Robert. "Tales of Terror from the State of Hidalgo." *Mexico Unexplained*, May 14, 2023.

Staff writer. "México Tales and Legends: The Charro Negro." *The Yucatán Times*, August 12, 2023.

Jalisco

Pint, John. "Mexico's Great Stone Balls, a Geological Attraction in the Hills of Jalisco." *Mexico News Daily*, January 4, 2019.

Pint, John. "The Mystery of Piedras Bola—How Did These Giant Rocks Form?" *Mexico News Daily*, September 23, 2022.

Staff writer. "The Ahuizotl." Mexicolore. Accessed May 12, 2024.

Vida, Melissa. "Legends of Jalisco: Devil's Bridge, Dragons & Blessings, Mysterious Stones." *The Guadalajara Reporter*, July 20, 2017.

Mexico City

Aguilar, Alan Gerardo Padilla. "The Origin of El Cucuy (El Coco)." *Latino Book Review*, May 14, 2023.

Baron d'Holbach, Paul-Henri Thiry. *Le Système de la Nature (The System of Nature)*. Library of Alexandria, 2012.

Becket, Stefan, and Tucker Reals. "Researcher Shows Bodies of Purported 'Non-Human' Beings to Mexican Congress at UFO Hearing." *CBS News*, September 13, 2023.

Bitto, Robert. "The Man-Bat of Northern Mexico." *Mexico Unexplained*, February 12, 2017.

de Sahagún, Bernardino. *The Florentine Codex*. University of Utah Press, 2002.

Golder, Joe, and Ryan Fahey. "Police Searching for NEW Loch Ness Monster in Depths of Reservoir by Massive Dam." *The Mirror*, March 21, 2023.

Janetsky, Megan. "Scientists Call Fraud on Supposed Extra-terrestrials Presented to Mexican Congress." *Associated Press*, September 13, 2023.

Proctor, Lucy. "Bogeymen: Five Scary Visitors in the Night." *BBC News*, December 24, 2012.

Michoacán

Clavijero, Francisco Javier. *Ancient History of Mexico*. Duke University Press, 1980.

Eberhart, George M. *Mysterious Creatures: A Guide to Crypto-zoology*. CFZ Press, 2002.

Staff writer. "The Zirahuen Lake Legend." *Latin Folk Tales*, May 4, 2016.

Morelos

Allen, Peter J., and Chas Saunders. "Azcatl." Godchecker, April 8, 2019.

Black, John. "Cipactli and Aztec Creation." *Ancient Origins*, May 19, 2013.

Staff writer. "Pelican Crossing—Not to Be Missed!" Mexicolore. Accessed March 12, 2024.

Nayarit

Coleman, Loren. *Mysterious America: The Revised Edition*. Paraview Press, 2001.

Shuker, Karl P. N. *Mystery Cats of the World Revisited: Blue Tigers, King Cheetahs, Black Cougars, Spotted Lions, and More*. Anomalist Books, 2020.

Staff writer. "Discovering Mexican Urban Legends: The Tale of La Quema de la Sierpe in Jala, Nayarit." *Nukari Quinta Boutique*, May 14, 2024.

Nuevo León

Binnall, Tim. "Goblin Filmed Scurrying Across Road in Mexico?" Coast to Coast AM, May 8, 2024.

Bitto, Robert. "Hombre Pájaro, Mexico's Bird Man Creature." *Mexico Unexplained*, September 18, 2022.

Bitto, Robert. "Legends from Nuevo León." *Mexico Unexplained*, January 30, 2022.

Staff writer. "Mothman Sightings in Latin America." *Espooky Tales*, February 15, 2022.

Oaxaca

Gil-Marin, Julieta. "Consider the Witch." *The Blackbird Review*, November 30, 2023.

Ubiera, Cheyenne R. "Fangs from the Dark: Disturbing Pictures of 'Shapeshifter' Revealed After Farmer Spots Man Who 'Got Powers from Aliens Morphing into Beast.'" *The US Sun*, November 23, 2023.

Vickery, Kirby. "'The Legend of the Bat' 2: The History." *Manzanillo Sun*, January 1, 2018.

Puebla

Bitto, Robert. "El Cuatlacas, the Other Mexican Bigfoot." *Mexico Unexplained*. March 26, 2018.

Bitto, Robert. "Seven Brief Legends from Puebla." *Mexico Unexplained*, December 23, 2019.

Mexico Unexplained. "Unveiling the Mysterious Cuatlacas: Mexico's Hidden Mountain Creatures." YouTube, January 26, 2024, 1:17. www.youtube.com/watch?v=GaReux46RNA.

Staff writer. "Haunted Churches." *Espooky Tales*, September 26, 2021.

Querétaro

Bitto, Robert. "Peña de Bernal, the Ultimate Mexican Paranormal Hotspot." *Mexico Unexplained*, August 22, 2021.

Mario Yair, T. S. "Quartz Grotto." *Atlas Obscura*, May 12, 2022.

Weibel, Barbara. "Magical, Mystical Peña de Bernal, Mexico." *Hole in the Donut Cultural Travel*, May 21, 2010.

Quintana Roo

Staff writer. "The Legend of Che Uinic, the Mythological Creature of the Yucatan Jungle." *El Universal*, March 13, 2021.

Wiener, James Blake. "Mayan Ocarina Depicting the Death God Ah." *World History Encyclopedia*, December 12, 2018.

San Luis Potosí

Reyes, Luis Ramirez. *Contact: Mexico: History of the UFO Phenomenon*. University of Texas, 1995.

Rosales, Albert. *Catalogue of Humanoid Cases 1965–2006*. Independently published, 2016.

Wayland, Tobias. "Resident of San Luis Potosí, Mexico, Claims to Have Seen Large, Thin Man 'Like a Giant.'" May 28, 2020. https://www.singularfortean.com.

Sinaloa

Fineman, Mark. "Tales of Bloodthirsty Beast Terrify Mexico." *The Los Angeles Times*, May 19, 1996.

Staff writer. "Legends of Sinaloa: The Appearance of the 'Calzonudo' of Jalpa, in Rosario." *The Mexican Post*, November 2, 2019.

Sonora

Holden, Will C. "Photos: 25 Alien-Like Skulls Discovered in Ancient Mexican Cemetery." KDVR-TV Denver, Colorado, December 21, 2012.

Murillo, Andrea. *They Are Among Us: Mimicry and Disappearance Against Reptilians in Mexico*. The College of Michoacán, 2023.

Tabasco

Bitto, Robert. "Legends from the State of Tabasco." *Mexico Unexplained*, November 26, 2023.

Callejas, Guillermo. "The Creation of Dtundtuncan and the Cenzontle." Forum-nepantla, April 19, 2020.

Davis, Whitney. "An Olmec 'Were-Jaguar' from the Yucatan Peninsula." *Cambridge University Press*, January 20, 2017.

Tamaulipas

Basner, Dave. "Naked Person Seen 'Running' on All Fours Like Animal in Security Footage." Z100 FM Portland, Oregon, December 12, 2023.

Bitto, Robert. "Tamaulipas Legends." *Mexico Unexplained*, June 6, 2021.

Tlaxcala

Campbell, Josianne Leah. "Tlahuelpuchi." *Erenow Philology Science Library*, April 30, 2024.

Mallett-Outtrim, Ryan. "Mexico's Monster Mountain: La Malinche." *Dissent! Sans Frontiéres*, June 27, 2016.

Veracruz

Browne, Malcolm W. "Researchers Call Rare Fish Evolutionary Link to Human Race." *New York Times*, April 14, 1991.

Naish, Darren. "It Had Wool, and Armour Plates, a Massive Beak, Horns, and It Smelled Veeeeery Bad: Whatever Happened to the Tecolutla Monster?" ScienceBlogs, July 9, 2008.

Yucatán

Michel, Diego. "10 Mythological Creatures of Mexico." *Vocal Media*. Accessed June 18, 2024.

Staff writer. "Cave Cow." *Encyclopedia of Cryptozoology*. Accessed June 20, 2024.

Zacatecas

Staff writer. "La Zona de Silencio (the Zone of Silence), and UFO Sightings." *Espooky Tales*, February 18, 2021.

Winick, Stephen. "La Llorona: An Introduction to the Weeping Woman." *Library of Congress*, October 13, 2021.

To Write to the Author

If you wish to contact the author or would like more information about this book, please write to the author in care of Llewellyn Worldwide Ltd. and we will forward your request. Both the author and the publisher appreciate hearing from you and learning of your enjoyment of this book and how it has helped you. Llewellyn Worldwide Ltd. cannot guarantee that every letter written to the author can be answered, but all will be forwarded. Please write to:

Jason Offutt
℅ Llewellyn Worldwide
2143 Wooddale Drive
Woodbury, MN 55125-2989

Please enclose a self-addressed stamped envelope for reply, or $1.00 to cover costs. If outside the U.S.A., enclose an international postal reply coupon.

Many of Llewellyn's authors have websites with additional information and resources. For more information, please visit our website at http://www.llewellyn.com.